Managing Quality

Managing Quality, 7th Edition, is a comprehensive and updated resource for mastering quality management in business. This new edition introduces timely coverage of organisational resilience, sustainability, risk management, agile development, and digital advances.

This book equips readers with the latest tools, techniques, and systems to ensure and sustain quality standards and practices in a rapidly evolving global market. It combines insights on strategic process improvement, operational excellence, and digital technologies, while offering new perspectives on resilience, risk management, and agile methodologies. With restructured material and updated terminology, it provides a clear and practical framework for understanding and implementing quality management principles effectively. Real-world case studies, problem-solving exercises, and examples from both large organisations and smaller medium enterprises (SMEs) ensure that readers can apply these concepts in diverse contexts, making it a valuable resource for both academic study and professional practice.

This book is ideal for professionals in operations, project management, and supply chain management, as well as students in business, applied science, and engineering disciplines.

The book is supported by online materials including exercises, questions, case studies, PowerPoint slides, as an instructor's manual. These tools provide a blended learning experience, making it an invaluable resource and support for educators, students, and professionals alike.

David Bamford, PhD, is Professor of Operations Management at the Manchester Metropolitan University. He is an experienced industrialist/academic with over 200 articles, book chapters, and reports, as well as having presented widely on his topics to both academic and practitioner audiences. Knowledge transfer has been at the centre of David's career strategy; it provides a focus and direction for his research, grant applications, publications, teaching, and major administrative roles. His research interests are quality management, operations management improvement strategies, supply chain management, sports operations management, strategic organisational change, and leadership.

Marina Papalexi, PhD, is Senior Lecturer in Operations and Project Management at Manchester Metropolitan University. Her research focuses on supply chain resilience, sustainability, digitalisation, and operations improvement strategies. She has published in high-impact international journals and contributed to several book chapters. Marina's work enhances academic literature by offering theoretical and empirical insights into complex delivery processes. Her research also spans quality management, sport management, and knowledge transfer. She has presented at international and national conferences, including EurOMA and BAM, and led research projects such as KTPs. Marina teaches BSc, MSc, and MBA courses in operations, supply chain, and project management.

Benjamin Dehe, PhD, is Associate Professor of Supply Chain Management at Auckland University of Technology (AUT). His research explores operational excellence, supply chain quality integration, innovation, and digital transformation, with a particular focus on Industry 4.0 technologies and decision-making processes. His work has been published in leading national and international journals and reflects a strong commitment to bridging theory and practice. Benjamin actively collaborates with organisations across manufacturing, construction, healthcare, and service sectors, contributing to the advancement of quality and supply chain management practices.

James Bamford, PhD, is Senior Lecturer in Project Management at York Business School, York St John University. He joined in 2021 after completing a PhD at the University of Huddersfield, where he also led undergraduate and postgraduate courses in operations, logistics, and project management. With a background in NHS strategic change projects across northern England, Jim brings extensive experience in healthcare and service management. He has led award-winning Knowledge Transfer Partnerships, receiving the ESRC Best Application of Social or Management Science and the TSB Best Partnership Award for innovation-led research. Jim's research interests include procurement, project management, and operations management.

Jon Tudor is the founder and owner of True North Excellence, a UK-based training and event company specialising in Lean Management and Operational Excellence. For over two decades he has been developing training and event programmes for organisations like the Manufacturing Institute and the Manufacturing Advisory Service and is an Advisory Board Member for the Society of Operational Excellence. He helped launch the Shingo Prize for Operational Excellence in the UK and has led manufacturing benchmarking study visits in the USA, UK, Europe, and Japan, visiting and studying over 300 companies.

Managing Quality

An Essential Guide and Resource Gateway

Seventh Edition

David Bamford, Marina Papalexi, Benjamin Dehe, James Bamford, and Jon Tudor

Routledge
Taylor & Francis Group
LONDON AND NEW YORK

Designed cover image: Shutterstock

Seventh edition published 2026
by Routledge
4 Park Square, Milton Park, Abingdon, Oxon, OX14 4RN

and by Routledge
605 Third Avenue, New York, NY 10158

Routledge is an imprint of the Taylor & Francis Group, an informa business

First edition published by Philip Allan 1990
Sixth edition published by Wiley 2016

British Library Cataloguing-in-Publication Data
A catalogue record for this book is available from the British Library

ISBN: 978-1-032-97584-9 (hbk)
ISBN: 978-1-032-97580-1 (pbk)
ISBN: 978-1-003-59436-9 (ebk)

DOI: 10.4324/9781003594369

Typeset in Times New Roman
by SPi Technologies India Pvt Ltd (Straive)

Access the Support Material: www.routledge.com/9781032975801

Contents

Acknowledgements

The authors would like to acknowledge the originator and founder of the Managing Quality series of textbooks, Barrie G. Dale. He is Professor Emeritus at Alliance Manchester Business School, and the first edition of Managing Quality was published in 1990. Barrie is the founder of the International Journal of Quality and Reliability Management and was its co-editor for over 25 years. He has authored several other books including, *The Road to Quality* (1993), *Quality Improvement through Standards* (2nd edition, 1994), *Managing Quality and Human Resources* (2nd edition, 1997), *Working in Partnership* (1998), *Managing Business Improvement and Quality* (1998), *Total Quality Management Blueprint* (1999), and *Quality Costing* (3rd edition, 1999).

Introduction to the Seventh Edition

This seventh edition of *Managing Quality: An Essential Guide and Resource Gateway* builds on the success of its predecessors, while embracing significant developments in the field of quality management.

In this edition, the term "quality management" is used in place of the previously emphasized "total quality management" (TQM). While TQM remains relevant in some domains and continues to appear in academic literature, the shift in terminology reflects the evolving language and practices in the field. The term "quality management" better captures the broader, more contemporary scope of the discipline, especially considering recent advances in digitalisation, agile methodologies, and organisational resilience.

This edition introduces a fully revised structure and a new team of co-authors, bringing fresh perspectives and expertise from both academia and industry. The book now includes timely coverage of topics such as risk management, agile development, and digital systems, ensuring it remains a vital resource for students, educators, and professionals navigating today's fast-changing global environment.

In the spirit of continuous improvement, all chapters have been thoroughly reviewed and updated. Some have been significantly revised or replaced to enhance clarity and relevance. The text is now organised into four parts:

1 The Development, Strategy, and Integration of Quality Management
2 The Business Context of Quality Management
3 Quality Management Tools, Techniques, and Systems
4 Quality Management Through Engagement

To support learning and teaching, a suite of online resources is now available, including teaching slides, additional case studies, and key questions for each chapter.

It is our hope that this new edition will continue to inspire readers to strengthen their commitment to quality and continuous improvement, equipping them with the tools and insights needed to thrive in a dynamic and demanding world.

The Authors, 2026

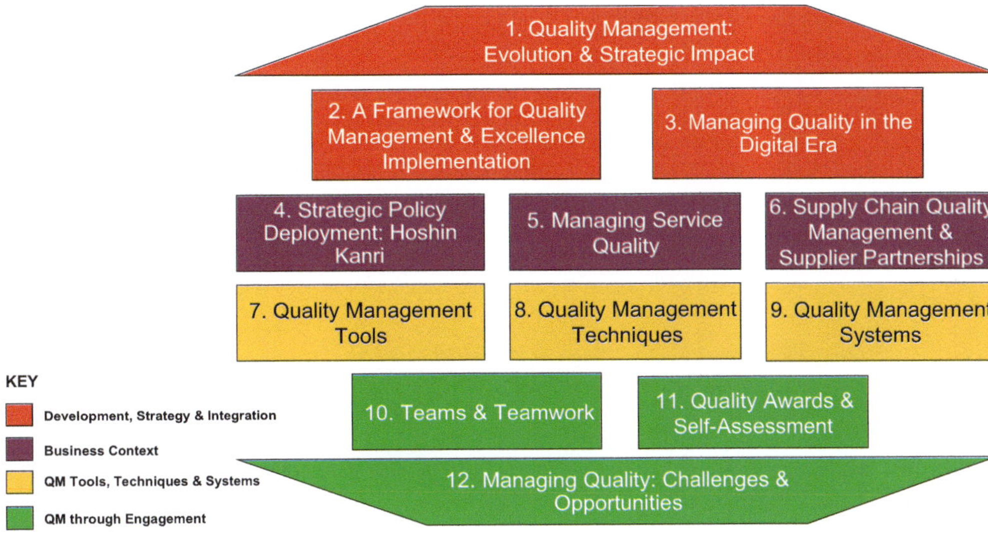

Figure I.1 Managing quality framework.

Source: Bamford, Papalexi, Dehe, Bamford, and Tudor.

Part 1

The Development, Strategy, and Integration of Quality Management

1 Quality Management

Evolution and Strategic Impact

Introduction

In today's global competitive marketplace, the demands of customers are gradually increasing as they anticipate constant services and product quality improvements. Satisfying customers' demand, and as a result, boosting competitiveness, remain primary strategic objectives for businesses. Managing quality plays a critical role in achieving these goals and serves as a fundamental driver of continuous improvement that leads towards better outcomes for customers and ultimately 'excellence'. Additionally, uncertainty, generated by disruptions, such as pandemics (Covid-19), the 'black swan' crisis, and other geopolitical influences, has highlighted the necessity of developing sustainability and resilience through innovative approaches (Tudor & Bamford, 2021; Bak & Papalexi, 2023). We argue that quality is key in reshaping the entire organisation model and logic (Enz et al., 2024). This presents an opportunity as well as a challenge for quality management practitioners. In some markets, although there is an increasing supply of competitively priced products and

DOI: 10.4324/9781003594369-2

services from lower labour cost countries, such as those in the Far East, Eastern Europe, China, Vietnam, and India, there is also limited application of quality management best practices (Sebastian-Coleman, 2022; King & Schwarzenbach, 2020). In fact, quality management and continuous improvement do not appear to have reached maturity in many developing economies (Brazil, Russia, India, China) (Andrade & Gonçalo, 2021; Persakis & Tsakalos, 2024). Continuous improvement in total business activities with a focus on the customer throughout the entire organisation and an emphasis on quality, sustainability, and resilience is one of the main means by which companies face up to these competitive threats. For this reason, many organisations are looking for quality management and continuous improvement to survive in increasingly aggressive markets and maintain a competitive edge over their rivals (Papalexi et al., 2020). As a result of the efforts made by organisations to respond to these marketplace demands, the quality of products, services, and processes has increased considerably during the last two decades. Oakland and Oakland (2020) state that:

> Total Quality has always been a key strategic factor for business success, but it is now more than ever required to compete successfully in the global markets of the twenty-first century.

Many organisations have had experiences with working on the transformation towards strategic quality process improvement using lean management or six sigma programmes, aiming at becoming more sustainable (achieving their financial success as well as fulfilling their environmental and social responsibility) and becoming more resilient to internal and external disruptions (Kumar et al., 2021; Frederico et al., 2023). According to Bamford et al., (2018) achieving and maintaining a quality culture is complex across all industrial sectors, but amplified based on particular industry characteristics. For example, quality management in the sports industry context is particularly interesting (Bamford et al., 2018).

In simple terms, quality management is defined in this book as the collective commitment and cooperation of everyone within an organisation, guided by a shared set of principles, to align all business processes in a manner that delivers products and services which provide value for money and aim to meet – and ideally exceed – customer needs and expectations.

Managing quality and organisational excellence involves ever-evolving practices of doing business in a bid to develop methods and processes that cannot be easily imitated by competitors, and enable the organisation to develop new capabilities of innovation and resilience.

This chapter provides an overview of quality management and continuous improvement. It opens by examining the different interpretations that are placed on the term 'quality'. It then examines why quality has grown in importance during the last decades. The evolution of quality management is described through the stages of inspection, quality control, quality assurance, and onwards to TQM. In presenting the details of this evolution, the drawbacks of a detection-based approach to quality are compared to the recommended approach of prevention. The chapter ends by presenting a summary of the points that organisations need to keep in mind when developing and advancing quality management and continuous improvement. This is done under the broad groupings of organising, systems and techniques, measurement and feedback, and changing the culture.

What Is Quality?

'Quality' has a variety of definitions, interpretations, and uses. Today, in a variety of situations, it is perhaps an overused word. Quality as a concept is quite difficult for many people to appreciate, and much confusion and myth surround it.

In a linguistic sense, quality originates from the Latin word 'qualis', which means 'such as the thing really is'. There is an international definition of quality: 'the degree to which a set of inherent characteristics fulfils requirements' (ISO 9001(2015)). However, in today's business world, there is no single accepted definition of quality. Irrespective of the context in which it is used, it is usually meant to distinguish one organisation, event, product, service, process, person, result, action, or communication from another.

Preventing confusion and ensuring that everyone in an organisation is focused on the same objectives, there should be an agreed definition of quality. For example, Betz Dearborn Ltd. defines quality as: 'That which gives complete customer satisfaction', and Rank Xerox (UK) as 'Providing our customers, internal and external, with products and services that fully satisfy their negotiated requirements'.

There are several ways or senses in which quality may be defined, some being broader than others, but they all can be boiled down to either meeting requirements and specifications or satisfying and delighting the customer. This has been expressed by the leading exponents and presented in Table 1.1 below.

Table 1.1 presents the early contributors to the quality management discipline and their work, which helps to understand the developments that have taken place in the quality management discipline. In the Western world the four best-known quality management experts are all Americans – Crosby (1979), Deming (1982), Feigenbaum (1961/1983) and Juran (1988). These four men have had a considerable influence on the development of TQM in organisations throughout the world. In addition to the approaches and philosophies of these four experts, the Japanese quality management culture is also widely publicised (Goetsch & Davis, 2014; Sallis, 2014). The work and ideas of a number of Japanese quality experts have been published in English. They include Imai (1986), Ishikawa (1985), Mizuno (1988), Nemoto (1987), Ozeki and Asaka (1990), Shingo (1986) and Taguchi, 1986. The ideas of these Japanese experts are all being applied in the West, but perhaps the work of Imai, Ishikawa, Shingo, and Taguchi is the best known. It is for this reason that it

Table 1.1 The received wisdom on quality management

American quality management experts	*Japanese quality management experts*
Philip Crosby (1979) 'Quality is Free' 'Cost of Quality'	Prof Kaoru Ishikawa (1985) 'Total Company Involvement'
Dr W Edwards Deming (1982) 'Quality is Management's Responsibility'	Genichi Taguchi (1986) 'Quality Engineering'
Armand Feigenbaum (1961/1983) 'Total Quality Control'	Musaaki Imai (1986) 'Kaizen – Continuous Improvement'
Dr Joseph M Juran (1988) 'Quality Trilogy – Planning, Control, Improvement'	Taiichi Ohno (1982) and Shigeo Shingo (1985) "Toyota Production System (TPS)"

is briefly reviewed here before considering some of the familiar concepts associated with how Japanese companies manage a process of continuous and company-wide improvement.

Qualitative View of Quality

When the word quality is used in a qualitative way, it is usually in a non-technical situation. ISO 9001(2015) says that 'the term "quality" can be used with adjectives such as poor, good or excellent'. Some examples related to this are:

- In advertising slogans to assist in building an image and persuade buyers that its production and services are the best: Quality is Job 1' – Ford Motor Company; Quality never goes out of style' – Levi's; 'Quality worth paying for' – John Deere; 'Quality in Everything We Do' – Ernst & Young; 'Committed to Quality, Committed to You' – Samsung; 'The Quality Goes In Before The Name Goes On' – Zenith; 'Quality comes as standard' – Siemens.
- By television and radio commentators (a quality player, a quality goal, a quality try).
- By directors and managers (quality performance, quality of communications).
- By people, in general (quality product, top quality, high quality, original quality, quality time, quality of communications, quality person, loss of quality, 100 per cent quality).

Quantitative View of Quality

The traditional quantitative term, which is still used in some situations, is acceptable quality level (AQL). This is defined in ISO 3951-2: 2013 as: 'the quality level that is the worst tolerable process fraction nonconforming when a continuing series of lots is submitted for acceptance sampling'. This is when quality is paradoxically defined in terms of non-conforming parts per hundred (i.e. some defined degree of imperfection).

An AQL is often imposed by a customer on its supplier in relation to a particular contract. In this type of situation, the customer will inspect the incoming batch according to the appropriate sampling scheme. If more than the allowed number of defects is found in the sample, the entire batch is returned to the supplier, or the supplier can, at the request of the customer, sort out the conforming from non-conforming products on the customer's site. The employment of an AQL is also used by some companies under the mistaken belief that trying to eliminate all defects is too costly.

The setting of an AQL by a company can work against a 'right-first-time' mentality in its people, as it appears to condone the production and delivery of non-conforming parts or services, suggesting that errors are acceptable to the organisation. It is tantamount to planning for failure. For example, take a final product that is made up of 3,000 parts: if the standard set is a 1 per cent AQL, this would mean that the product is planned to contain 30 non-conforming parts. In reality, there are likely to be many more because of the vagaries of the sampling used in the plan or scheme, whereby acceptance or rejection of the batch of product is decided.

Another example of a quantitative measure is to measure processes using sigmas (a sigma is a statistical indication of variation) and defects per million opportunities (DPMO). A sigma is essentially a measuring device that is an indication of how good a product or service is. The higher the sigma value, the lower the number of defects. For example, 3 sigma equals 66,807 DPMO, while 6 sigma equals 3.4 DPMO (these values assume a normal distribution with a process shift of 1.5 sigma). The sigma level is a means of calibrating performance in relation to customer needs. Six Sigma (a quality improvement framework and programme) has used sigmas to improve productivity and quality and reduce costs. Six sigma is the pursuit of perfection and represents a complete way of tackling process improvement from a quantitative approach, involving many of the concepts, systems, tools, and techniques described in this book. The six-sigma concept is very popular as a business improvement approach. The key features include a significant training commitment in statistics and statistical tools, problem-solving methodology and framework, project management, a team-based project environment, people who can successfully carry out improvement projects (these are known as black belts and green belts, based on the martial arts hierarchy), leaders (master black belts), and project champions.

Uniformity of the Product or Service Characteristics Around a Nominal or Target Value

Figure 1.1 presents the inside/outside specification dilemma; only the product or service dimensions that are within the design specification or tolerance limits can be considered acceptable. The difference between what is considered to be just inside or just outside the specification is marginal. It may also be questioned whether this step change between pass and fail has any scientific basis and validity.

Designers often establish specification limits without sufficient knowledge of the process by which the product and/or service is to be produced/delivered and its capabilities. It is often the case that designers cannot agree amongst themselves about the tolerances/specifications to be allocated, and they tend to establish a tighter tolerance than is justified to provide safeguards and protect themselves. In many situations there is inadequate communication on this matter between the design and operation functions. Fortunately, this is changing with the increasing use of simultaneous or concurrent engineering and more recently the use of Industry 4.0 technologies (Bibby & Dehe, 2018), such as virtual reality, 3D-printing, big data analytics (Dehe et al., 2025), and artificial intelligence. Indeed, within footwear manufacturing, efforts to design a digital supply chain were challenging due to difficulties in communication between stakeholders. However, the adoption of virtual reality proved to be a powerful tool as it reimagined the prototyping process. Specifically, it minimised the production of physical samples by replacing them with 3D models presented in a virtual environment. As a result, communication between designers and producers was enhanced, information accuracy improved, errors were minimised, and most importantly, lead time was reduced dramatically from 90 days to 10 days. The main issue of working to

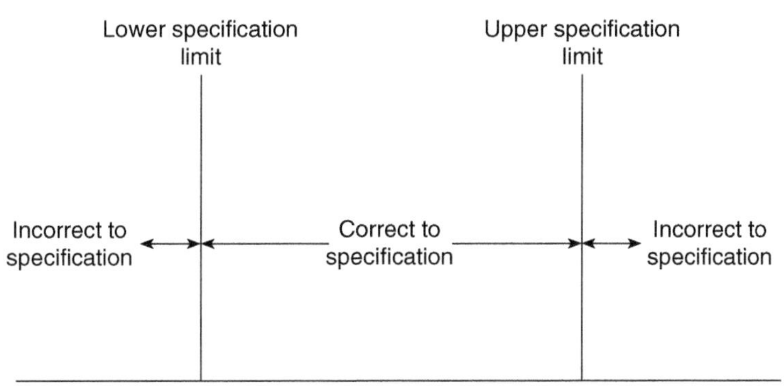

Figure 1.1 The inside/outside specification dilemma.

the specification limits is that it frequently leads to tolerance stack-up; for example, in a manufacturing situation, parts may not fit together correctly at the assembly stage. This is especially the case when one part, which is just inside the lower specification limit, is assembled to one that is just inside the upper specification. If the process is controlled, such that a part is produced around the nominal or a target dimension with limited variation (see Figure 1.2), this problem does not occur, and the correctness of fit and smooth operation of the final assembly and/or end product is enhanced.

The idea of reducing the variation of part characteristics and process parameters so that they are centred around a target value can be attributed to Taguchi (1986). He wrote that the quality of a product is the (minimum) loss imparted by the product to society from the time the product is shipped. Among the losses he includes are time and money spent by customers, consumers' dissatisfaction, warranty costs, repair costs, wasted natural resources, loss of reputation, and, ultimately, loss of market share.

The relationship between design specification and variation of the process can be quantified by a capability index, for example, C_p which is a process potential capability index:

$$C_p = \frac{\text{Total specification width}}{\text{Process variation width}}$$

Conformance to Agreed and Fully Understood Requirements

This concept is attributed to Crosby (1979). He believed that quality is not comparative and that there is no such thing as high quality or low quality, or quality in terms of goodness, feel, excellence, and luxury. In other words, quality is an attribute (a characteristic which, by comparison to a standard or reference point, is judged to be correct or incorrect), not a

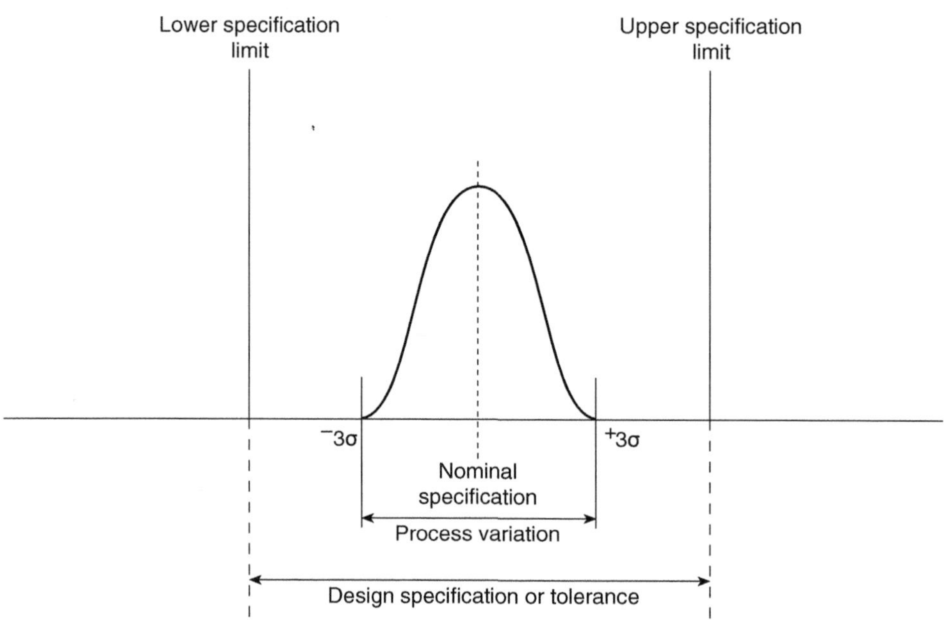

Figure 1.2 Design tolerance and process variation relationship.

variable (a characteristic which is measurable). Crosby made the point that the requirements are all the actions required to produce a product and/or deliver a service that meets the customer's expectations, and that it is management's responsibility to ensure that adequate requirements are created and specified within the organisation.

Fitness for Purpose/Use

Juran (1988) was the first to use this definition of quality. He classifies 'fitness for purpose/use' into the categories of: quality of design, quality of conformance, abilities, and field service. Focusing on fitness for use helps to prevent the over-specification of products and services. Over-specification can add greatly to costs and tends to militate against a right-first-time performance.

Satisfying Customer Expectations and Understanding Their Needs and Future Requirements

Satisfying customers and creating customer enthusiasm through understanding their needs and future requirements is the crux of quality management and continuous improvement. Managing quality is all about customer orientation, and many company missions are based entirely on satisfying customer perceptions. Customer requirements for quality are increasing and becoming stricter. There are increasing levels of intolerance of poor-quality goods and services and low levels of customer service and care. Additionally, there is an increasing awareness of and demand for sustainable and ethical products and services that have informed and changed the quality standards (Elmor et al., 2025). In most situations, customers have a choice: they are not willing to jeopardise their own business interest out of loyalty to a supplier who does not perform as they expected, and they will simply go to a competitor. In the public sector the customer may not have this choice; however, they can go to litigation, write letters of complaint, cause disruption, and use elections to vote officials out of office.

Superior-performing organisations go beyond satisfying their customers: they emphasise the need to delight them by giving them more than what is required in the contract. These organisations create a total experience for their customers, which is unique in relation to the offerings of competitors (which is called 'the experience economy', see Pine and Gilmore (2011)). The wisdom of this can be clearly understood considering the situation where a supplier has given more than the customer expected (e.g. an extra glass of wine on an aircraft; a sales assistant going out of their way to be courteous and helpful, and providing very detailed information), and the warm feelings generated by this type of action.

A customer-focused organisation also puts considerable effort into anticipating the future expectations of its customers (i.e. surprising quality), by working with them in long-term relationships, helping them to define their future needs and expectations. Their aim is to build quality into the product, service, system, and/or process as upstream as is practicable. Excitement and loyalty are the words used to describe this situation.

A mechanism for facilitating a continuous two-way flow of information between themselves and their customers is considered necessary. There is also a variety of means available to companies for them to assess issues such as:

- How well the brand is respected
- How well they are meeting customer expectations

- What customers' chief causes of concern are
- What the main complaints are
- What suggestions for improvements customers might have
- How well they act on what the customer says
- How they might add value to the product and/or service
- What the best means of differentiating themselves in the marketplace are

Organisations tend to focus on increasing the level of contact with the customer and becoming more transparent (Kotcharin et al., 2024). 'All you need is trust': examining trust, information sharing, and supplier performance perception in SMEs from a social exchange perspective. (Cogent Business & Management, 11(1), 2415523.) These 'moments of truth' (Carlzon, 1987, also see Fatma, 2014) occur far more frequently in commerce, public organisations, and service-type situations than in manufacturing organisations. They use the following practices to increase the level of customer contact:

- Webinars
- Online communities and forums
- Using 'test' consumers and mystery shoppers
- Focus groups
- Customer interviews
- Market research
- Dealer information
- Questionnaire surveys
- Product reports
- Trailing the service and/or product
- Trade shows

Customer complaints are one indication of customer satisfaction, and many organisations have several metrics measuring such complaints. BS ISO 10002 (2018) provides guidance on how to develop an effective complaints management system to analyse and use complaints effectively. The rationale is that managing complaints in a positive manner can enhance customer perceptions of an organisation, increase lifetime sales and values, and provide valuable market intelligence.

Why Is Quality Important?

To explore why quality is important, consider a time when a customer experienced poor product or service quality. Reflect on how it made them feel, the actions they took in response, and the people they shared the experience and its outcome with. From an operations perspective, quality can be conceptualised as meeting the customers' expectations; however, understanding the customers' perspective of quality remains inherently complex (Slack et al., 2022). This complexity stems from the multifaceted nature of customer expectations, which are fundamentally shaped based on their individual knowledge, experience, and socio-cultural contextual factors. While one person might perceive quality through the lens of cost-effectiveness, another may prioritise sustainability or product durability.

The following presents quantitative evidence in relation to these arguments (Dale et al., 2016):

- Problems decrease customer loyalty by 15 per cent to 30 per cent
- 50 per cent of individual consumers and 25 per cent of business customers who have problems never complain to anyone at the company
- If the call centre can resolve a customer's problem using quality service, thus changing a dissatisfied customer to a satisfied one, the company usually gets an increase in loyalty of 50 percentage points
- Market leaders can charge between 5 per cent and 10 per cent premiums for outstanding quality and service

Quality Is Not Negotiable

An order, contract, or customer, which is lost on the grounds of non-conforming product and/or service quality, is much harder to regain than one lost on price or delivery terms. In several cases the customer could be lost forever; in simple terms, the organisation has been outsold by the competition.

If one has any doubt about the truth of this statement, just consider the number of organisations that have gone out of business or lost a significant share of a market, and consider the reported reasons for them getting into that position. Quality is one of the factors that is not negotiable, and in today's business world, the penalties for unsatisfactory product quality and poor service are likely to be punitive.

Quality Is All-Pervasive

There are a few single-focus business initiatives that an organisation may deploy to increase profit. Quality management and continuous improvement encompass not only enhancements to products, services, and processes, but also improvements related to cost efficiency, productivity, and the engagement and development of people. A number of surveys show that customers are willing to pay more for improved quality of products and services, such as: according to a survey by Deloitte (2023), 30 per cent of respondents stopped using certain products due to sustainability-related concerns, and 26 per cent are willing to pay more for sustainable products, which is associated with higher quality production methods. McKinsey & Company's Consumer Sentiment Survey discovered that approximately 70% of consumers would pay a premium for products they perceive as having exceptional quality.

Managers sometimes say that they do not have the time and resources to ensure that product and/or service quality is done right the first time. They go on to argue that if their people concentrate on planning for quality, then they will be losing valuable operational time, and consequently, output will be lost, and costs will rise. Despite this argument, management and their staff will make the time to rework the product and service a second or even a third time and spend considerable time and organisational resources on corrective action and placating customers who have been affected by the non-conformances.

Remember – 'There is never enough time to do it right, but there's always time to do it over' (attributed to John W Bergman).

Quality Means Improved Business Performance

Winners of the Deming Application Prize (a prize awarded to companies for their effective implementation of company-wide quality philosophies, for details see Chapter 11), had financial performance in terms of earning rate, productivity, growth rate, liquidity, and net worth that was above the average for their industries. The Malcolm Baldrige National Quality Award (MBNQA) reported similar impact across different industry sectors, such as: manufacturing, health care, service, education and small business; this programme was established to enhance the competitiveness of U.S. businesses based on the seven criteria of: leadership; strategic planning; customer focus; measurement, analysis, and knowledge management; workforce focus; operations focus; and results.

Similarly, the European Foundation Quality Management (EFQM) Excellence Model, which was developed based on MBNQA, has been used by over 50,000 organisations across Europe, and is now global. The Canadian Award, which was developed based on the National Quality Institute's Framework for Organisational Excellence, has been designed to support continuous quality improvement for nonprofit organisations, such as government, education, and health care.

Cao et al. (2024) highlighted that the three best-known quality awards globally are the Baldrige Award, the Deming Prize, and the EFQM Award, and underlined that they have served as global benchmarks for organisational excellence and contributed to advancing the quality evolution.

The Cost of Non-Quality Is High

Based on a variety of companies, industries, and situations, the cost of quality (or to be more precise, the cost of not getting it right the first time) ranges from 60 to 90 per cent of an organisation's total cost of quality, according to Crosby; see Bamford et al. (2023) for details. Indeed, Ali et al. (2021) found that the cost of quality of a multinational company in Tangier, specialised in wiring, was 14.5% of the turnover, including 10.4% of non-quality. An organisation should compare its profit-to-sales turnover ratio to that of its quality costs-to-sales turnover ratio in order to gain an indication of the importance of product and service quality to corporate profitability.

Chiarini (2020) examined the impact of the ISO 9001 non-conformity process on the cost of poor quality in different sectors, including: chemical, pharmaceutical, mechanical, food, ceramic, and steel. They found that the ISO 9001 non-conformity process has the same impact on these six different sectors, and highlighted that the reduction in cost of poor quality was no more than 27.14 per cent. They suggested that other important factors could reduce the total cost of poor quality, including the adoption of improvement techniques such as Six Sigma and TQM.

Customer Is King

In today's markets, customer requirements are becoming increasingly more rigorous, and their expectations of the product and/or service in terms of conformance, reliability, dependability, durability, interchangeability, performance, features, appearance, sustainability, serviceability, user-friendliness, safety, and environmental friendliness are also increasing. These days, many superior-performing companies talk in terms of being 'customer-centred'. At the same time, it is likely that the competition will also be improving, and, in addition, new and low-cost competitors may emerge in the marketplace. Consequently, there is a need for continuous improvement in all operations of a business, involving everyone in the company. The organisation, which claims that it has achieved high quality and organisational excellence, will be overtaken by the competition. Once the process of continuous improvement has been halted, under the mistaken belief that high quality has been achieved, it is much harder to restart and gain the initiative on the competition (see Figure 1.3).

Quality Is a Way of Life

Quality is a way of organisational and everyday life. It is a way of doing business, living and conducting one's personal affairs. Quality is driven by a person's own internal mechanisms – 'heart and soul', 'personal beliefs'.

An organisation committed to quality needs the quality of working life of its people in terms of participation, involvement, and development, and the quality of its systems, processes, and products.

The seven quality management principles are defined in ISO 9001 (2015) as:

- *Customer focus*. The primary focus of quality management is to meet customer requirements and to strive to exceed customer expectations.
- *Leadership*. Leaders at all levels establish unity of purpose and direction and create conditions in which people are engaged in achieving the quality objectives of the organisation.
- *Engagement of People*. Competent, empowered, and engaged people throughout the organisation enhance its capability to create value.
- *Process approach*. Consistent and predictable results are achieved more effectively and efficiently when activities are understood and managed as interrelated processes that function as a coherent system.
- *Improvement*. Successful organisations have an ongoing focus on improvement.
- *Evidence-based Decision Making*. Decisions based on the analysis and evaluation of data and information are more likely to produce desired results.
- *Relationship Management*. For sustained success, organisations manage their relationships with interested parties, such as suppliers.

The Evolution of Quality Management

Systems for improving and managing quality have evolved rapidly in recent years. During the last three decades or so, simple inspection activities have been replaced or supplemented

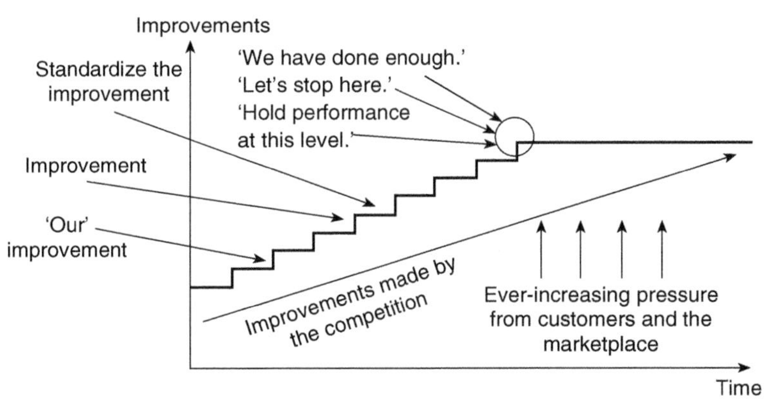

Figure 1.3 Quality improvement: a continuous process.

by quality control, quality assurance has been developed and refined, and now many companies, using a process of continuous and company-wide improvement, are working towards 'excellence'.

The term Total Quality Management (TQM) was first introduced in 1969 and established in the 1980s, where companies adopted the TQM guidelines and emphasised company-wide quality. Within this 21-century review, it is more common for this to be known as 'Quality Management' (QM) or 'Organisational Excellence'. In this progression, four discrete stages can be identified: inspection, quality control, quality assurance, and QM; it should be noted that the terms are used here to indicate levels in a hierarchical progression towards quality management/organisational excellence (Figure 1.4).

Inspection

Conformity evaluation by observation and judgement accompanied as appropriate by measurement, testing, or gauging. (ISO 9000 (2015))

At one time inspection was thought to be the only way of ensuring quality, the 'degree to which a set of inherent characteristics fulfils requirements' (ISO 9000 (2015)). Under a simple inspection-based system, one or more characteristics of a product, service, or activity are examined, measured, tested, or assessed and compared with specified requirements to assess conformity with a specification or performance standard. In a manufacturing environment, the system is applied to incoming goods and materials, manufactured components, and assemblies at appropriate points in the process and before finished goods are

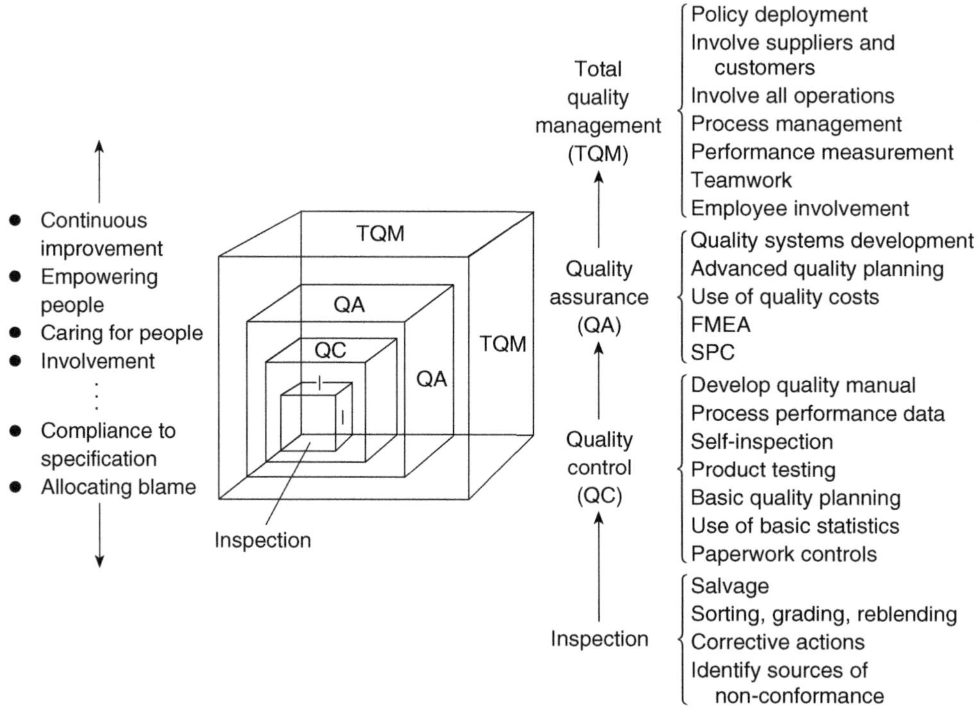

Figure 1.4 The four levels in the evolution towards QM/Organisational Excellence.

passed into the warehouse. In service, commercial, and public service-type situations, the system is also applied at key points, sometimes called appraisal points, in the production and delivery processes. The inspection activity is, in the main, carried out by dedicated staff employed specifically for the purpose, or by self-inspection of those responsible for a process. Materials, components, paperwork, forms, products, and goods that do not conform to specifications may be scrapped, reworked, modified, or passed on concession. In some cases, inspection is used to grade the finished product, as, for example, in the production of cultured pearls. The system is an after-the-event screening process with no prevention content other than, perhaps, identification of suppliers, operations, or workers who are producing non-conforming products/services. There is an emphasis on reactive, quick-fix, corrective actions, and the thinking is department-based. Simple inspection-based systems are usually wholly in-house and do not directly involve suppliers or customers in any integrated way.

Quality Control

> Part of quality management focused on fulfilling quality requirements.
>
> (ISO 9000 (2015))

Under a system of quality control one might expect, for example, to find in place detailed product and performance specifications, a paperwork and procedures control system, raw material and intermediate-stage product-testing and reporting activities, logging of elementary process performance data, and feedback of process information to appropriate personnel and suppliers. With quality control there will have been some development from the basic inspection activity in terms of sophistication of methods and systems, self-inspection by approved operators, use of information, and the tools and techniques which are employed. While the main mechanism for preventing off-specification products and services from being delivered to customers is screening inspection, quality control measures lead to greater process control and a lower incidence of non-conformance.

Those organisations whose approach to the management of quality is based on inspection and quality control are operating in a detection-type mode.

What Is Detection?

In a detection or 'firefighting' environment, the emphasis is on the product, procedures, and/or service deliverables and the downstream producing and delivery processes; it is about getting rid of the bad things after they have taken place. Considerable effort is expended on after-the-event inspecting, troubleshooting, checking, and testing of the product and/or service, and providing reactive 'quick fixes' in a bid to ensure that only conforming products and services are delivered to the customer. In this approach, there is a lack of creative and systematic work activity, with planning and improvements being neglected and defects being identified late in the process, with all the financial implications of this in terms of the working capital employed. Detection will not improve quality but only highlight when it is not present, and sometimes it does not even manage to do this. Problems in the process are not removed but contained and are likely to come back. It also leads to the belief that non-conformances are due to the product/service not being inspected enough and that operators, not the system, are the sole cause of the problem.

With a detection approach to quality, non-conforming 'products' (products are considered in their widest sense) are culled, sorted, and graded, and decisions are made on concessions, rework, etc. It is not unusual to find products going through this cycle more than once. While a detection-type system may prevent non-conforming products, services, and paperwork from being delivered to the customer (internal or external), it does not prevent them from being made. Indeed, it is questionable whether such a system does, in fact, find and remove all non-conforming products and services. Physical and mental fatigue decreases the efficiency of inspection, and it is commonly claimed that, at best, 100 per cent inspection is only 80 per cent effective. It is often found that with a detection approach, the customer also inspects the incoming product/service; thus, the customer becomes a part of the organisation's quality control system.

In this type of approach, a non-conforming product must be made, and a service delivered before the process can be adjusted; this is inherently inefficient in that it creates waste in all its various forms: all the action is 'after-the-event' and backward-looking. The emphasis is on 'today's events', with little attempt to learn from the lessons of the current problem or crisis. It should not be forgotten that the scrap, rework, etc., are extra efforts, and represent costs over and above what has been budgeted, and which ultimately will result in a reduction of bottom-line profit.

An environment in which the emphasis is on making good non-conformance rather than preventing it from arising in the first place is not ideal for engendering team spirit, cooperation, and a good climate for work. The focus tends to be on switching the blame to others, people making themselves 'fireproof', not being prepared to accept responsibility and ownership, and taking disciplinary action against people who make mistakes. In general, this behaviour and attitude emanate from middle management and quickly spread downwards through all levels of the organisational hierarchy.

Organisations operating in a defensive manner are often preoccupied with the survival of their business and little concerned with making improvements.

Quality Assurance

Finding and solving a problem after a non-conformance has been created is not an effective route towards eliminating the root cause of a problem. A lasting and continuous improvement in quality can only be achieved by directing organisational efforts towards planning and preventing problems from occurring at source. This concept leads to the third stage of QM development, which is quality assurance:

> Part of quality management focused on providing confidence that quality requirements will be fulfilled.
>
> (ISO 9000 (2015))

Examples of additional features acquired when progressing from quality control to quality assurance are, for example, a comprehensive QM system to increase uniformity and conformity, use of the seven quality control tools (histogram, check sheet, Pareto analysis, cause-and-effect diagram, graphs, control chart and scatter diagram), statistical process control, failure mode and effects analysis (FMEA), and the gathering and use of quality costs. Above all one would expect to see a shift in emphasis from mere detection towards prevention of non-conformances. In short, more emphasis is placed on advanced quality planning, training, critical problem-solving tasks, improving the design of the product, process, and services, improving control over the process, and involving and motivating people.

What Is Prevention?

Quality assurance is a prevention-based system that improves product and service quality, and increases productivity by placing the emphasis on product, service, and process design. By concentrating on source activities and integrating quality into the planning and design stage, it stops non-conforming products from being produced or non-conforming services from being delivered in the first place; even when defects occur, they are identified early in the process. This is a proactive approach compared with detection, which is reactive. There is a clear change of emphasis from downstream to the upstream processes and from product to process; 'product out' to 'customer in'. This change of emphasis can also be considered in terms of the plan, do, check, act (PDCA) cycle. In the detection approach the 'act' part of the cycle is limited, resulting in an incomplete cycle, whereas, with prevention, act is an essential part of individuals and teams striving for continuous improvement as part of their everyday work activities.

With prevention there is a clearly defined feedback loop with both negative and positive feedback into the process, product, and service development system.

Quality is created in the design stage and not at the later control stage; the majority of quality-related problems are caused by poor or unsuitable designs of products and processes. In the prevention approach, there is a recognition of the process as defined by its input of people, machines, materials, method, management, and environment. It also brings a clearer and deeper sense of responsibility for quality and eliminates the root cause of waste and non-value-adding activity to those actually producing and delivering the product and/or service.

Changing from detection to prevention requires not just the use of a set of tools and techniques, but the development of a new operating philosophy and approach, which requires a change in management style and way of thinking. It requires the various departments and functions to work and act together in cross-functional teams to discover the root cause of problems and pursue their elimination. Quality planning and continuous improvement truly begin when top management includes prevention as opposed to detection in its organisational policy and objectives and starts to integrate the improvement efforts of various departments. This leads to the next level, that of QM.

Quality Management/Organisational Excellence

The fourth level – QM/organisational excellence – involves the application of quality management principles to all aspects of the organisation, including customers and suppliers, and their integration with the key business processes. It is a company-wide approach to quality, with improvements undertaken on a continuous basis by everyone in the organisation. Individual systems, procedures, and requirements may be no higher than for a quality assurance level of QM, but they will pervade every person, activity, and function of the organisation. It will, however, require a broadening of outlook and skills and an increase in creative activities from those required at the quality assurance level. The spread of the QM philosophy would also be expected to be accompanied by greater sophistication in the application of tools and techniques, increased emphasis on people (the so-called soft aspects of QM), process management, improved training and personal development, and greater efforts to eliminate wastage and non-value-adding activities. The process will extend beyond the organisation to include partnerships with suppliers and customers and all stakeholders of the business. Activities will be reoriented to focus on the customer, internal, and external, with the aim to build partnerships and go beyond satisfying the customer to delighting them. The need to self-assess progress towards business excellence is also a key issue.

There are many interpretations and definitions of QM. Put simply, QM is the cooperation of everyone in an organisation and associated business processes to produce value-for-money products and services which meet and hopefully exceed the needs and expectations of customers. QM is both a philosophy and a set of guiding principles for managing an organisation to the benefit of all stakeholders.

Although QM has been a significant influence on organisations' perspective regarding quality through introducing the company-wide responsibility for quality, there are other equally important models and methods that have contributed to managing quality. Figure 1.5 presents and briefly describes them.

Figure 1.5 The models and methods supporting quality management in businesses.

The Concept of Excellence

The concepts of excellence and world-class organisations represent an evolution beyond traditional QM. While QM focuses on continuous improvement and the involvement of all employees in quality processes, excellence refers to achieving outstanding, sustainable results across all areas of performance. World-class organisations demonstrate exceptional capabilities not only in quality but also in leadership, strategy, innovation, sustainability, and stakeholder engagement. These organisations benchmark against global best practices and strive for long-term value creation. Excellence frameworks, such as EFQM, Shigo, or the Malcolm Baldrige, provide structured approaches to assess and guide progress towards world-class performance; these will be discussed further in Chapter 11.

Summary

This chapter defines quality, highlighting its importance. It discusses the cost of the lack of quality throughout the production of a product and/or the delivery of a service. It presents the early contributors of the QM discipline and their work, which helps to understand the developments that have taken place in the QM discipline. It very importantly describes the evolution of quality management, highlighting the role of key elements, such as inspection, quality control, detection, quality assurance, and the key models and methods that support QM in businesses.

References

Ali, M.B., Mazini, D.B., Ghailani, I., & Barrijal, S. (2021). Evaluation of quality cost in a multinational company located in Tangier, Morocco. *International Journal of Productivity and Quality Management*, 32(2), 216–245.

Andrade, C.R.D., & Gonçalo, C.R. (2021). Digital transformation by enabling strategic capabilities in the context of "BRICS". *Revista de Gestão*, 28(4), 297–315.

Bak, O. and Papalexi, M. (2023). A preliminary analysis of blockchain impact on sustainable supply chains: COVID-19 perspective. In *Industry 4.0 Technologies: Sustainable Manufacturing Supply Chains: Volume II-Methods for transition and trends* (pp. 133–150). Singapore: Springer Nature Singapore.

Bamford, D., Forrester, P., & Reid, I. (2023). *Essential guide to operations management: Concepts & case notes*, 2nd edition, Routledge.

Bamford, D., Hannibal, C., Kauppi, K. and Dehe, B. (2018). Sports operations management: examining the relationship between environmental uncertainty and quality management orientation. *European Sport Management Quarterly*, 18(5), 563–582.

Bibby, L., & Dehe, B. (2018). Defining and assessing industry 4.0 maturity levels–case of the defence sector. *Production Planning & Control*, 29(12), 1030–1043.

BS ISO 10002 (2018). *Quality management. Customer satisfaction. Guidelines for complaints handling in organizations*. London: British Standards Institution.

Cao, Y., Peng, X., & Prybutok, V. (2024). Achieving sustainable excellence: A theoretical comparison of established quality awards. *Quality Management Journal*, 31(2), 138–153.

Carlzon, J. (1987), *The moments of trust*. Cambridge, Mass.: Ballinger.

Chiarini, A. (2020). Effect of ISO 9001 non-conformity process on cost of poor quality in capital-intensive sectors. *International Journal of Quality & Reliability Management*, 32(2), 144–155.

Crosby, P.B. (1979). *Quality is free*. New York: McGraw-Hill.

Dale, B., Bamford, D., van der Wiele, T. (2016). *Managing Quality: An essential guide and resource gateway*, 6th edition. UK: John Wiley & Sons.

Dehe, B., Subasinghage, M., & Mirzaei, M. (2025). Orchestrating resources for Big data analytics implementation in manufacturing SMEs: insights into managerial role and engagement. *Production Planning & Control*, 1–33.

Deloitte (2023) High prices continue to come at a cost.... https://www.deloitte.com/uk/en/about/press-room/high-prices-continue-to-come-at-a-cost-to-the-planet.html

Deming, W.E. (1982). *Quality, productivity and competitive position.* Cambridge, Mass.: MIT, Centre of Advanced Engineering Study.

Elmor, L., Ramos, G.A., Vieites, Y., Andretti, B., & Andrade, E.B. (2025). Environmental sustainability considerations (or lack thereof) in consumer decision making. *International Journal of Research in Marketing*, 42(4), 1203–1228.

Enz, M.G., Ruel, S., Zsidisin, G.A., Penagos, P., Bernard Bracy, J., & Jarzębowski, S., (2024). Supply chain strategies in response to a black-swan event: A comparison of USA, French and Polish firms. *The International Journal of Logistics Management*, 35, 1–32.

Fatma, S. (2014). Antecedents and consequences of customer experience management a literature review and research Agenda. *International Journal of Business and Commerce*, 3(6), 32–49.

Feigenbaum, A.V. (1961/1983). *Total quality control.* 1st/3rd edn. New York: McGraw-Hill Inc.,US.

Frederico, G.F., Kumar, V., Garza-Reyes, J.A., Kumar, A., & Agrawal, R. (2023). Impact of I4. 0 technologies and their interoperability on performance: Future pathways for supply chain resilience post-COVID-19. *The International Journal of Logistics Management*, 34(4), 1020–1049.

Goetsch, D., & Davis, S.B. (2014). *Quality management for organizational excellence: Introduction to Total Quality Managemnet*, Vol. 427, 6th edition. Upper Saddle River, NJ: Pearson.

Imai, M. (1986). *Kaizen: The Key to Japan's competitive success.* New York: Random House Business Division.

Ishikawa, K. (1985). *What is total quality control? The Japanese way*, trans. D.J. Lu. Prentice Hall.

ISO 9000 (2015). Quality management systems: Fundamentals and vocabulary. Available from website: http://www.iso.org

ISO 9001 (2015). Seven Principles Of Quality Management As Per ISO 9001:2015 Committee Draft. Available from website: http://www.iso.org

Juran, J.M. (editor-in-chief) (1988), *Quality control handbook.* New York: McGraw-Hill.

King, T., & Schwarzenbach, J. (2020). *Managing data quality: A practical guide.* UK: BCS Publishing.

Kotcharin, S., Dehe, B., & Boonchoo, P. (2024). 'All you need is trust': examining trust, information sharing, and supplier performance perception in SMEs from a social exchange perspective. *Cogent Business & Management*, 11(1), 2415523.

Kumar, V., Ramachandran, D., & Kumar, B., (2021). Influence of new-age technologies on marketing: A research agenda. *Journal of Business Research*, 125, 864–877.

Mizuno, S. (1988). *Company-wide total quality control.* Tokyo: Asian Productivity Organization.

Nemoto, M. (1987). *Total quality control for management.* Englewood Cliffs, NJ: Prentice Hall.

Oakland, J.S., & Oakland, M.A. (2020). *Total quality management and operational excellence: text with cases*, 5th edition. Routledge: New York.

Ohno, T. (1982). How the Toyota production system was created. *Japanese Economic Studies*, 10(4), 83–101.

Ozeki, K., & Asaka, T. (1990). *Handbook of quality tools.* Buckinghamshire: Productivity Europe.

Papalexi, M., Bamford, D., & Breen, L. (2020). Key sources of operational inefficiency in the Pharmaceutical Supply Chain. *Supply Chain Management: An International Journal*, 25(6), 617–635.

Persakis, A., & Tsakalos, I. (2024). Navigating the Storm: How Economic Uncertainty Shapes Audit Quality in BRICS Nations Amid CEO Power Dynamics. *Journal of Risk and Financial Management*, 17.

Pine, B., & Gilmore, J. (2011). *The experience economy.* Boston, Mass.: Harvard Business Review Press.

Sallis, E. (2014). *Total quality management in education.* 3rd edition. London: Routledge.

Sebastian-Coleman, L. (2022). *Meeting the challenges of data quality management.* USA: Academic Press.

Shingo, S. (1985). *A Revolution in Manufacturing: the SMED System (Translation Book).* UK: *Productivity Press.*

Shingo, S. (1986) *Zero Quality Control: Source Inspection and the Poka-Yoke System.* Cambridge, Mass.: Productivity Press.

Slack, N., Brandon-Jones, A., & Burgess, N. (2022). *Operations management*, 10th edition. UK, Pearson Education, Limited.

Taguchi, G. (1986). *Introduction to quality engineering.* New York: Asian Productivity Organization.

Tudor, J., & Bamford, D. (2021). How COVID-19 has transformed the way manufacturers operate. https://www.mmu.ac.uk/news-and-events/news/story/14457

2 A Framework for Quality Management and Excellence Implementation

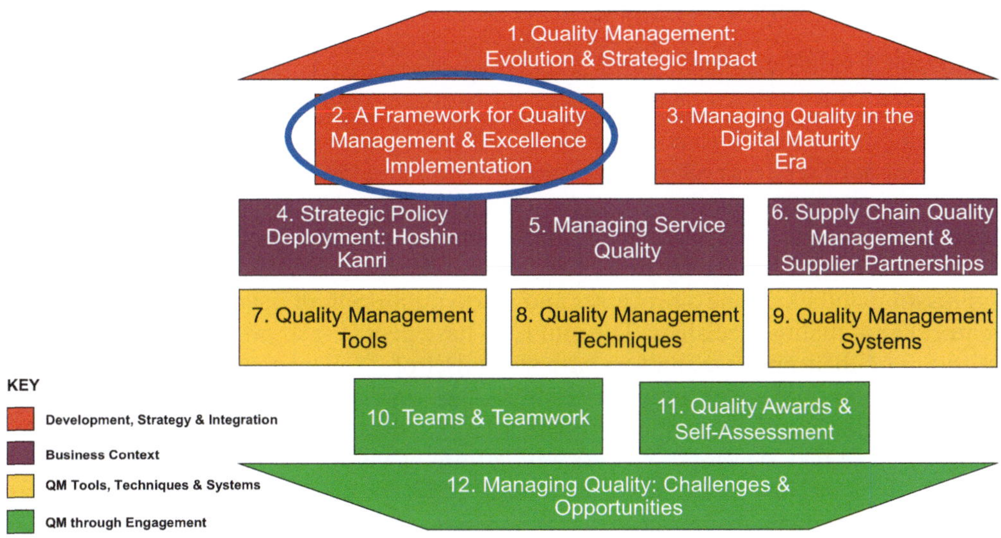

Introduction

There are several approaches that can be followed in the introduction of quality management QM and continuous improvement (CI) (e.g. Kassela et al., 2017; Papalexi et al., 2020, 2022; Dehe et al., 2022; Bambrick et al., 2025). These include:

1 The deployment of principles and practices based on a set of guidelines provided by established tools/techniques (e.g. lean and six sigma).
2 A prescriptive step-by-step approach.
3 Methods outlining the wisdom, philosophies and recommendations of internationally respected experts on the subject or a company production system (e.g. Crosby, Deming, Feigenbaum and Juran, or Toyota Production System (TPS)).
4 Self-assessment and excellence Prizes such as the Deming Prize, the Shingo Model, the Malcolm Baldrige model or the EFQM model (see Chapter 11).

DOI: 10.4324/9781003594369-3

With all this available advice and prescription, it is not surprising that there is sometimes inertia on the part of senior management teams who are faced with the task of introducing a culture of QM and CI in their organisations.

It is up to the management team of each organisation to identify the approach that best suits their needs and business operations. Indeed, it is not unusual for an organisation to find that its QM approach is not working out as planned and to switch to another approach. Some of the main ways of deploying QM and CI are examined in this chapter. However, importantly, this chapter presents the QM, CI and excellence culture model that can guide organisations, companies and leadership teams on the key elements of quality management, including: commitment and leadership, planning and organisation, tools and techniques, education and training, employee involvement, teamwork, measurement and feedback, and culture change.

Change and Continuous Improvement

Changing the life-long behaviour, customs, and practices of an organisation is not easy. Organisations committed to quality and CI will strive permanently to improve the quality of their goods or services and processes and are committed to change (Papalexi et al., 2020; Dehe et al., 2022). The decision to start working on QM and CI is often taken by the senior management team. They must encourage a total corporate commitment to continually improve every aspect of the business. Quality is an integral part of the management of an organisation and its business processes and is too important an issue to delegate it only to technical and quality specialists. The aim is to have people take ownership of the quality assurance of their processes and to have a mindset of CI. This situation is not a natural phenomenon and does not happen overnight, and senior managers must be prepared to spend time coaching people along this path and providing the necessary influences.

To start and then develop a process and culture of CI, an infrastructure is required to support the associated tasks and departments. people need to be able to devote time to quality planning, and to prevention and improvement activities (Bamford et al., 2023). Mong et al., (2021) point out that CI is directly related to maintenance performance. Organisation's strengths and opportunities for improvement in relation to the management of quality need to be identified. This typically takes the form of an internal assessment of employees' views and perceptions (internal and group assessments, and questionnaire surveys), a systems audit, a cost of quality analysis, and obtaining the views of customers (including those accounts which have been lost) and suppliers about the organisation's performance in terms of product, service, people, administration, innovation, strengths and weaknesses. This type of internal and external assessment of perspectives should be carried out on a regular basis to gauge the progress being made towards QM and CI and help decide the next steps.

In most companies, the major motivations for introducing or reinitiating QM and CI include:

- Environmental, national and business factors and changing circumstances such as a global crisis (e.g. Covid-19 or financial crisis), slow economic growth or severe competition.
- A lack of effective long-range planning.
- A need to develop new products that are attractive to the marketplace.
- Slow growth in sales and market, leading to stagnation of the business.

- Concerns about how to achieve the long-term plan of the organisation.
- Complacency about current profits and a failure to recognise the seriousness of the situation.
- Organisational business weaknesses such as:
  - Lack of advanced planning for quality
  - Lack of liaison between development, design and manufacturing departments
  - Lack of alignment between the management policies throughout the organisation
  - A poor approach to the solution of problems
  - The morale of workers was poor
  - Chronic defects in the manufacturing process
  - Problems at operational start-up due to insufficient pre-planning.

Arkadan et al. (2024) argued that competition acts as a catalyst, setting off a chain reaction, which enhances quality awareness in the market, resulting in demanding customers and chief executives behaving as external and internal change agents, respectively see Figure 3.1.

Forces for Change

The Role of Leadership Towards QM and CI Culture

Many writers, about quality management, agree that unless the senior leadership team plays an active role and takes the lead to improve quality within an organisation, attempts and gains made by individuals and departments will be short-lived. The limited success usually surfaces in a lack of empowerment at lower levels of the organisational hierarchy. Senior managers should be sensitive to the fact that some employees will resist the change to QM and CI. The usual reasons for this are that they are uncertain of the nature and impact of QM and CI, and their ability to cope. If senior managers are personally involved in the change process, it can help to break down these barriers.

Mohammad et al. (2011) reported that supportive leadership, consistent support of top management, and employee involvement are critical to QM and CI success. One of the key roles of senior management is to develop effective strategies to support and enhance the chances of achieving business excellence (Dehe et al., 2022; Humeisat, 2022).

They need to create and promote an environment in which, for example:

- People can work together as a team, and teamwork becomes an integral part of business activities.
- Effective two-way communication is in place.
- People are involved in the business through decision-making.
- People improve on a continuous basis the processes under their control.
- People direct their attention to identifying, satisfying, delighting and winning over customers, whether they be internal or external.
- Ideas are actively sought from everyone.
- Mistakes are freely admitted without recriminations and are perceived as an opportunity for improvement (i.e. a 'blame-free' culture).
- Recognition is given for improvement activities.

However, change is not something that any department or individual takes too easily, and administering changes in organisational practices must be considered with care. Communications, regarding the importance of the establishment of a QM and CI approach within organisations, are essential. This can be done by getting involved in activities such as:

- Setting up and chairing a QM and CI steering group or quality council.
- Identifying the major quality issues facing the organisation and becoming personally involved in investigating them.
- Getting involved in quality planning, audit, improvement meetings and organisational housekeeping.
- Leading and/or attending quality training courses.
- Organising and chairing defect review and customer groups.
- Instigating and carrying out regular audits, self-assessment and diagnosis of the state of the art of QM and CI.
- Dealing with customer complaints and visiting customers and suppliers.
- Leading customer workshops, panels and focus groups.
- Visiting, on a regular basis, all areas and functions of the business, and discussing improvement issues.
- Developing, communicating and then following a personal improvement action plan.

Thus, other factors must also be present, of which market pressure (e.g. intense competition and demanding customers) has by far the greatest impact.

Competition

Competition is fierce in today's business environment, and quality is recognised as a key consideration in many purchasing decisions. There is little doubt that quality is an essential part of the marketing mix as companies seek ways to effectively differentiate their products and/or services from those of their competitors. Many successful companies now advertise their products and/or services based on quality and reliability rather than price. For example, consider the number of advertisements in which the word quality is featured. They also frame quality using different terms such as transparency, sustainability or even resilience. There is a great discussion in supply chain management literature regarding opportunities and approaches to enhance transparency, traceability and visibility within multi-tiered supply networks to increase reputation, competitiveness and customer satisfaction (Chen, 2022; Papalexi et al., 2022; Budler et al., 2023; Bambrick et al., 2025; Mirzaei et al., 2025). Indeed, the motivation for improvement is provided here by the need to stay competitive, and the change agent is the customer whose awareness of quality has been enhanced. As a result of such pressures, suppliers of goods and services have themselves become demanding customers and seek improved levels of quality conformance from their own suppliers.

Demanding Customers

Demanding customers with high product and service quality expectations and an established reputation for quality can be very effective change agents. In addition to providing tangible evidence of the value of reputation and standing, they have the potential for bringing about radical and permanent changes in attitudes towards CI among their suppliers through the requirements they place on them. For example, as awareness of social and environmental sustainability – that includes employees' treatment, ethical sources and products, and waste minimisation – increases, consumers put massive pressure on companies to adopt sustainable production processes and develop transparent networks. Therefore, many major purchasers have policies, which outline what is required of suppliers in terms of their approach to quality management (e.g. the QS 9000 quality system standard; fair trade and modern slavery statements). Documents such as these describe fundamentals that must be incorporated into a supplier's quality planning methods and quality system to control and improve quality. Each supplier is responsible for building on these fundamentals to develop an effective quality system, and products and services which are defect-free (Bambrick et al., 2025). Without this evidence, it is often the case that companies cannot enter a supply network, which has a massive impact upon their production process (e.g. limited sources). Many quality-conscious purchasers assess and evaluate supplier performance. Some also provide resources to help suppliers implement tools, techniques and systems, improve workplace layout, and give guidance on problem-solving.

'Fresh-Start' – Environmental Factors Situations

The degree of entrenchment of attitudes, and hence the difficulty of changing them, is related to the length of time an organisation has been established, its size, staff turnover rate, managerial mobility, markets, competitors, and many other factors which influence the 'performance' of an organisation. A fresh-start situation, therefore, provides excellent opportunities to make rapid, fundamental changes to attitudes and relationships.

A 'greenfield' venture may be the setting up of a new company, a new operational direction for an existing company (e.g. creation of a new strategic business unit as part of a diversification programme), an established company relocating to a new factory, or a company establishing a new operation in existing premises after rationalising plant, product lines, manpower, etc. A greenfield venture provides an opportunity for the introduction of a process of CI in a situation where there is no prior history of lame excuses, acceptance of non-conformance, shipping non-conforming products to meet production targets, providing and accepting a poor level of service and poor delivery performance. In a greenfield venture, there is an opportunity to start from scratch without any vested interests or inhibiting procedures to overcome. It is an opportunity for senior management to try to do all the things that should be done to engender a culture of CI.

How Do Companies Get Started?

Once change has been triggered, organisations need to translate enhanced quality awareness and organisational need for improvement into effective action. At this stage an organisation's senior management team ask questions such as:

- What should we do?
- What are the priorities?
- What advice do we need?
- Should the approach be top-down or bottom-up?
- Do we need an umbrella term? Do we have to use the terms QM, lean or six sigma? What are the alternative methodologies and approaches to business excellence?
- How do we apply the tools and techniques?
- How quickly should we proceed?

- What courses and training should we attend?
- Which network of companies should we attempt to join?
- What packages and programmes should we buy?

Their dilemma is often compounded not just by a lack of knowledge of QM and CI but also by challenges in managing and leading organisational change. The overwhelming quantity and variety of available advice, which is often conflicting, sometimes biased and sometimes incorrect and misdirected, simply adds to the confusion and chaos. It is not surprising then that there is sometimes inertia on the part of senior management teams who are faced with the task of introducing a formal process of CI in their organisations.

Approaches to QM and CI

Each writer about QM and organisational excellence develops and outlines an approach which reflects their own background, values and experience. These approaches include:

1 Methods outlining the wisdom, philosophies and recommendations of the internationally respected experts or company on the subject.
2 Prescriptive step-by-step approaches.
3 A listing of QM principles, which are presented in the form of an implementation plan and a set of guidelines or Non-prescriptive methods in the form of a framework or model.
4 Self-assessment methods based on the excellence models on which the Deming Prize, the Shingo Model, the Malcolm Baldrige National Quality Award (MBNQA) and the European Quality Award (EQA) are based.

Thus, there are a number of ways to get started, and it is up to each organisation to identify the approach which best suits its needs and business operation.

Applying the Wisdom of the Quality Management Experts and World-Class Organisations

The writings and teachings of Crosby, Deming, Feigenbaum and Juran or even from a company like Toyota, are a sensible starting point for any organisation introducing QM and CI. An unusual approach is for an organisation to adopt the teachings of one of these quality management experts and/or world-class organisations and attempt to follow their programme. The argument for this approach is that each expert and/or company has a package which works, which gives some form of security, it provides a coherent framework, gives a discipline to the process and provides a common language, understanding and method of communication.

Applying a Consultancy Package

Some companies decide to adopt the programme of one of the major management consultancies on the grounds that it is a self-contained package, which can be suitably customised for application throughout their organisation. A number of the consultancy packages are based on the teachings and wisdom of the quality management experts like Crosby, Juran and Deming.

It is important for a company to understand that the use of a consultant organisation does not relieve the senior management team of their own responsibilities for QM and CI; it is their responsibility to own the improvement process and to exercise leadership. A key part of consultancy is the transfer of skills, technology and knowledge, and when the project is complete, the training and guidance provided by the consultant must remain within the organisation in order that the QM and CI process and culture can progress and develop. They also help to change the way managers think and behave. The consultant should be perceived by the organisation as an 'implementation' mechanism and not as an initiator of QM and CI. Management consultancies bring their expertise and skills to the company and provide the resources, experience, disciplines, objectivity and the catalyst for getting the process started. An organisation can benefit from the use of this expertise if it is willing to accept the recommendations of the consultants.

The QM, CI and Excellence Culture Framework

In this seventh edition of our textbook, we adopt the Dynamic Capability Theory (DCT) (Teece, 2007) to explain the capabilities that organisations must develop to achieve quality, CI and eventually excellence and 'world-class' recognition. By adopting a DCT lens, we can shed some light on how organisations can implement QM and CI tools, techniques and systems to reconfigure their resources, develop capabilities to increase their performance and develop a competitive edge. Specifically, the skills and competencies of sensing, seizing, and reconfiguring are considered dynamic capabilities that are fundamental determinants of firm performance. These dynamic capabilities help organisations swiftly address potential risks by detecting environmental changes (sensing), determining the necessary resources and actions to navigate uncertainties, implementing adaptable operations (seizing), and reorganising resources to restore and maintain operational continuity (reconfiguring). This theory facilitates the examination of a company's ability to restructure systems and frameworks in response to unpredictable business environments and generate opportunities (Bruyaka et al., 2024).

Importantly, the development of QM and CI capabilities and especially maintaining operational continuity requires organisations to change their culture (Papalexi et al., 2020). Changing the culture is a key element in QM and CI and has wide-ranging implications for the whole organisation; it requires the introduction and acceptance of individual, group and organisational change (Bamford et al., 2024). QM and CI provide the opportunity to make and influence behaviours and attitudes which have real effects on internal and external relationships and the way the organisation conducts its business. The change of culture must be planned to avoid ambiguity and facilitate improvement; a significant number of academic articles have focused on this element (Brown, 2011; Green, 2012; Maroofi, 2012; Gimenez-Espin et al., 2013; Valmohammadi & Roshanzamir, 2015; Palumbo & Douglas, 2023) (Figure 2.1).

Sensing Dynamic Capability – Organising

As mentioned earlier in the chapter, the key influences for a change are derived from the competition, demanding customers and environmental factors; therefore, the role of leadership is crucial here to recognise those changes and develop a clear strategy and capabilities to support organisations throughout an adaptation process. The main organising

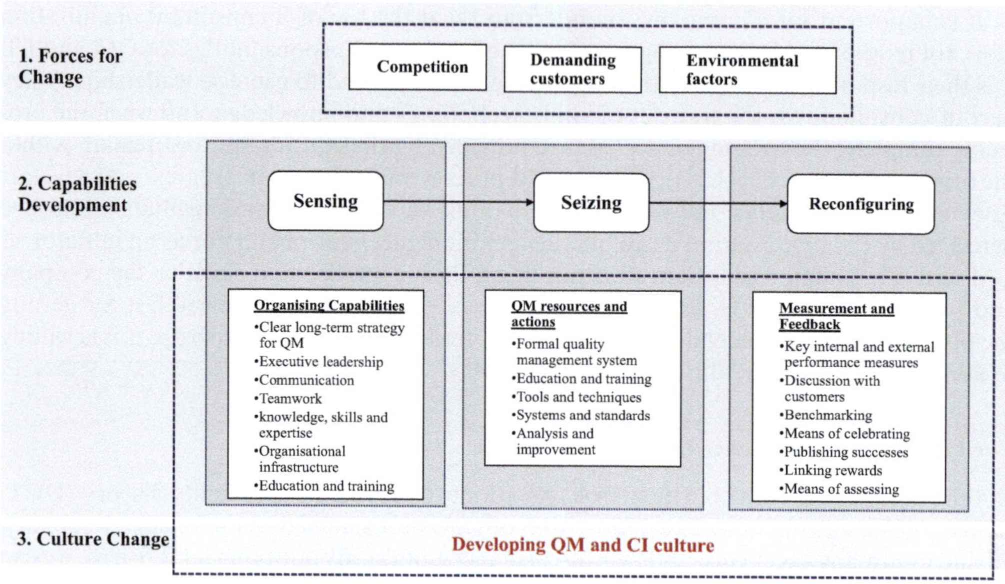

Figure 2.1 Quality Management, Continuous Improvement and Excellence Cultural Model.

capabilities that organisations need to develop to deal with any quality-related issue during a change are listed below:

Organising Capabilities

- A clear long-term strategy for quality, CI and excellence should be formulated and integrated with other key business strategies, departmental policies, organisational culture and objectives. This also includes the development of a quality policy and quality strategy. The aim should be to integrate them with the long-term plans of the business. Chapter 4 covers these aspects.
- Executive leadership, tangible commitment and support should be recognised as being a crucial capability. Such commitment should be demonstrated in actions such as allocating time to understanding and involvement in QM and CI, being visible and accessible ('management by wandering about'), holding discussions with people at the operating level of the business, providing words of encouragement and advice, 'quality' placed at the top of every business meeting agenda, identification of key performance measurements, use of tools and techniques in their everyday work activities, developing personal action plans, seeking feedback on their style of management, acting as a mentor to improvement teams, attending training sessions, writing articles about QM, CI and the concept of excellence in the company newsletter, ensuring that any decisions made are consistent with the agreed plans and objectives, and exhibiting a passion for quality.
- Communication capability is critical at this stage. Vision and mission statements, which are concise and understandable to all employees, should be developed, displayed and communicated in company-unique language. It is also important to outline what needs to be done to make these statements and the associated company values become a reality, including the benefits that will accrue from a culture of quality and excellence and how it will affect the way employees go about their jobs.

- Teamwork should be established and become part of the organisation's method of working. In the first place it is suggested that a review is undertaken of any teams which are already established, in conjunction with their previous and current projects. Following this, task forces/project teams and cross-functional improvement teams should be established to address the major problems facing the organisation, followed by the setting up of departmental improvement teams.
- The knowledge, skills and expertise of the stakeholders (internal and external). Useful expertise is often available within the organisation. Such people know the internal culture of the organisation, its processes and the unique problems. This expertise should not be overlooked. It is always beneficial to combine internal expertise with external consultants' knowledge and skills.
- An organisational infrastructure should be established, which will ultimately facilitate local ownership of a quality and excellence culture. Direction should be provided by the QM steering group, but the time it takes for people to accept such ownership for quality and CI should not be underestimated. Actions include deciding the membership of the group, role and meeting frequency, setting up, as appropriate, local steering groups, identification of improvement coordinator (full-time or part-time), facilitators and team leaders, along with clear definitions of their roles, ensuring how the actions developed by improvement teams can be carried through and agreeing on budgets.
- A formal programme of education and training should be established. This is important to build the skills of employees, and should involve basic job skills and process training, including induction, quality and excellence awareness, customer care, and training in the use of tools, techniques and systems.

Seizing Dynamic Capability – QM and CI Resources and Actions

- This pillar of the framework involves the implementation of the quality management and CI resources and actions to provide the necessary controls and discipline, and the standardisation of improvements. It involves the deployment of quality management tools, techniques and principles, for example, developing quality at the source, capturing the 'voices' of customers, focusing on data-driven decision-making, eliminating process variation, achieving incremental improvements, problem-solving through root cause analysis and involving everyone. Key elements at this stage include:
- The tools and techniques applicable at different stages of the improvement process should be identified; these are often linked to lean and six sigma methodologies. In the first place, consideration should be given to identifying which tools and techniques employees are familiar with and those which are in regular use. Tools and techniques should be classified as core and optional, depending on their nature and impact and the environment (e.g. manufacturing or service) in which they are being applied.
- The right type of training targeted at the right people should be developed; it should emphasise the why and how of the tools and techniques and the benefits of their use. When tools, techniques and systems have been deployed incorrectly, an additional set of problems in the introduction of QM is created. Suitable training packages on tools and techniques should be developed and customised for the organisation – this is perceived to be very important in some situations (e.g. public services).
- The use of a formal quality management system should be considered if one is not in place. If such a system is already in use, then some evaluation of its contribution to QM

is vital; the objective should be to continually improve and strengthen the quality system and ensure that any improvements are built into the system.

- Any other systems and standards which may be required as part of future contractual or legislative requirements, or simply in order to compete in certain markets, should be identified and implemented (e.g. ISO9001). If relevant systems and standards are integrated with the improvement initiative, it is less likely that the organisation will have conflicting priorities and policies, and confusion will be reduced.
- Process analysis and improvement should be a continual part of the organisation's improvement process. There should be a focus on processes (e.g. business planning and control and order generation) rather than functions within the organisation. Process analysis and innovation give emphasis to the centrality of quality throughout the business process and focus attention on customer and supplier relationships.

Reconfiguring Dynamic Capability – Measurement and Feedback

This final element of the framework suggests that organisations must develop CI capabilities that allow them to reorganise resources to restore and maintain operational continuity. To achieve this, organisations need to collect data from different sources that enable them to measure their performance and identify the areas that need to be further developed. This pillar of the framework enables the 'voice of the customer' to be translated into measures of performance with which the organisation can identify, and on which it can improve. It also deals with internal measures of performance, supplier assessment and development and rewards and recognition. Key actions at this stage include:

- Key internal and external performance measures should be identified to assess the progress being made with the quality culture developed, and to ensure that customers are satisfied. The measurement process involves a two-way flow of information between the organisation and its customers and suppliers, and these parties should be consulted as part of the process of deciding what measurements to make. However, it should be accepted from the outset that measuring customer satisfaction can be difficult and painful. Care must be taken to ensure that appropriate measures are developed, defined clearly and used. The chosen performance measures, in line with the organisation's balanced scorecard, should help to facilitate the integration of quality into the business processes of the organisation and encourage all employees to participate. It is important to build results and corrective actions into improvement plans and standardise the improvements across the organisation.
- Discussion with customers (internal as well as external) about the performance expected and their needs and expectations should be undertaken, using a variety of techniques. This must be an ongoing exercise to ensure that gaps between actual performance and customer needs and expectations are identified and analysed, and actions are put in place to close the gap. In going about this exercise, it is also important to assess the relationship between the sales and marketing functions and the strengths of each. The main objective of all this is to build a partnership with customers and to develop customer loyalty to build a competitive advantage.

Issues that Must Be Considered in this Marketplace Research Include

- How well the organisation is meeting customer expectations
- How customers perceive they are treated
- The main complaints from customers
- Suggestions the customer might have for improvements and what else may be required in terms of products, services and features
- How the organisation rates against the competition
- Whether the data which has been collected is used to generate improvements which benefit the customer.

In some organisations it may be necessary to initiate suitable systems for identifying customer needs. Customers must also be encouraged and invited to challenge the organisation which is delivering the product or service. The trend is for increasing the level of contact with customers (internal and external), and such 'moments of truth' occur far more frequently in commerce, public organisations and service-type situations than in manufacturing organisations.

Systems to Identify Customer Needs Include

- Customer workshops
- Client service and call centres
- Panels and clinics
- Focus groups
- Customer interviews
- Market research
- Surveys: mail (including electronic), telephone, comment cards, point of purchase (survey designs should vary in length, contact and format)
- Trailing the service and/or product
- Field trials of new products
- Using 'test' consumers and mystery shoppers
- Feedback from professional and trade associations
- Product launches
- Field contacts

Often potential sources of information are customers lost, and customers gained, the data which the finance and accounts department holds on customers, and field failures and warranty claims. There must be a methodology and system for analysing and feeding back the data gathered from customers by such means (i.e. customer service measurement); the same applies to data on competitors. A considerable number of researchers have focused on using the receiving feedback to understand customer needs (i.e. Dienst et al., 2014; Celuch et al., 2015; de Haan, Verhoef & Wiesel, 2015; Gremyr et al., 2022; Okeke et al., 2024).

- Benchmarking should be considered once the organisation has taken some steps to improve quality. The benchmarking of a small number of strategic processes helps employees to see the need for change and thereby gives impetus to the improvement process. The concept of benchmarking is a proven technique for assisting companies with a process of CI. It is a process whereby internal performance and practices are compared to those of other companies, including the superior-performing ones, in a bid to develop, improve and achieve the best practice that leads to superior performance (see Parast & Adams, 2012; Dehe et al., 2022).

- Means of celebrating and communicating success should be considered, and methods developed for recognising the efforts of teams and individuals. The issue of ownership of QM is linked to providing adequate recognition, rewards and incentives for quality efforts, and in this way, the message that quality is a strategic concern is reinforced. Two quotes worthy of mention are 'What gets measured gets done' (Anders Scharp, former CEO, Electrolux) and 'What gets rewarded gets repeated' (Anne Van't Haaff, former corporate quality manager, KLM).
- Publishing successes is an effective means of communicating how people have tackled improvements. Recognition and communication of success can be facilitated in a number of ways such as quality news-sheets, team briefs, quality action days, team competition/celebration days, quality conferences, presentations by the president and/or CEO, supplier award days, 'thank you' notes, small tokens of appreciation such as mugs, pens, meals, certificates and trophies, publicity in the company newsletter and so on. Personal 'thank you' and 'praise' notes from senior management are often seen as a more genuine recognition than buying people with money.
- Linking rewards to improvement activities and results must be considered, although it is controversial. Financial payment for participation in improvement activities, in particular, those schemes relating to individuals, should be discouraged but perhaps not overlooked. CI should be a natural part of every person's job, but people at different levels of an organisation have widely differing expectations of what improvement means to them personally and to the company.
- Means of assessing the progress of the business towards world-class performance should be used. For example, the Shingo Prize, the MBNQA criteria for performance and the EFQM excellence model should be considered (see Chapter 11).

Changing the Culture

It is not possible to identify key actions for this stage, but there are several features which should be considered:

1 A self-assessment, from both management and employee perspectives, of the status of the organisational culture should be undertaken before firm plans for change are developed. Senior management must be prepared to resolve conflicts and overcome resistance to change.

2 Change must be recognised as ongoing, rather than as a prerequisite to the introduction of quality and strategic process improvement culture. Some degree of culture change in terms of senior management commitment and leadership and provision of adequate resources must, however, take place prior to and as part of the organising stage. For example, the effective use of tools and techniques, developing the quality management system to meet the requirements of the ISO 9000 series, teamwork, the impact of successful improvement projects, presentations, recognition, effective channels of communication, etc., are all activities which can contribute to culture change. There are, of course, other activities which will contribute to the culture change process, which may not connect directly with quality and process improvement.

3 Change should be planned and take place in a consistent and incremental manner. Experience indicates that if the change is too great and unplanned, the organisation will revert to the status quo. The planned changes must be outlined in specific terms and, where possible, qualified against a timescale. Employee attitude surveys, customer surveys and internal buyer–supplier workshops are also useful for identifying culture change indicators. Examples of possible changes include:

   - Delegate decision-making and the responsibility for taking actions down to the lowest possible level and spread the power base.
   - Teach managers to adopt a listening, consulting and learning style of leadership.
   - Train managers to act as trainers.
   - Provide opportunities for management to listen to the views of staff and customers and develop a listening and learning style of leadership.
   - Enable every employee to visit a customer and other parts of the business.
   - Introduce the concept of associates rather than employees.
   - Introducing cross-functional team activity.
   - Recognise and respect people's contributions to the business.
   - Change to a cellular type of organisation.

   The role of people within the organisation should be recognised. The way that they are treated is vital, since they are an intellectual asset whose value to the organisation can be increased by careful nurturing or decreased by poor management. It should also be recognised that most organisations are made up of people of differing ages, backgrounds, skills, abilities, levels of enthusiasm, levels of flexibility and ability to accept change. If culture change is to be successful, these people-based factors must be considered.

4 Teamwork is an important facilitator in culture change, but organisations must ensure that the organisational infrastructure can adapt to the changes which teamwork will bring. The operating characteristics of the teams to be employed in QM should be defined and communicated (see Chapter 11). It is also essential that participants in teams and other improvement activities are volunteers, not 'conscripts'.

5 The interrelationship of all activities in the organisation, and the way in which they contribute to the overall quality of service and product provided, should be identified, so that conflict is minimised and QM becomes part of the way in which the business is run. Such conflict typically arises at the middle management level, where the impact of strategic initiatives meets the problems of day-to-day running of the organisation. In any large organisation there will be a variety of initiatives going on at one time, many of which will affect staff directly (e.g. development of information technology, introduction of Manufacturing Execution System (MES), cost-cutting exercises, and marketing promotions), and these may indirectly contribute to the quality of product and service provided.

6 Factors which indicate that quality and excellence have started to change culture should be identified. Without such factors it is difficult to know whether cultural change is taking place, and the concept may be undermined by 'lack of results'. Factors that indicate that culture is changing include:

   - Changes to procedures and systems are easier to make.
   - Motivators and champions start to emerge from various parts of the organisation.
   - People talk about processes and not functions.
   - People are not afraid of expressing their views.

- People show a positive response to recognition.
- Employees are viewed by senior management as an asset and not a cost.
- People volunteer to take on tasks which would previously have involved considerable negotiation between management and unions.
- Ideas and suggestions start to flow from the shop floor.
- Team meetings scheduled outside of the team shift, without pay.
- CI goes on in the face of organisational instability.

Summary

This chapter has argued that a formal approach to QM and CI is triggered by one or more of four factors, namely the CEO, competition, demanding customers and fresh-start, environmental factors, and situations.

The point has been made that there is no 'right' way of introducing and developing QM and CI culture. There are several approaches, and these have been examined in the chapter. It is senior management's responsibility to select the approach which best suits its business and operating environment and any constraints which may exist. Employees must be involved and understand the transformation process. The approach should always be tailored to the organisation, and 'off-the-shelf' packages should be avoided. Senior management has much to gain by networking with their counterparts in different businesses, and this exchange of ideas and concerns and discussion of common issues can help to fine-tune the approach which is being used and to advance the development of QM and CI.

Finally, this chapter presents the quality and excellence culture, which supports organisations in developing the capabilities required to establish and maintain a culture towards excellence.

References

Arkadan, F., Macdonald, E.K., & Wilson, H.N. (2024). Customer experience orientation: Conceptual model, propositions, and research directions. *Journal of the Academy of Marketing Science*, 52, 1560–1584.

Bambrick, S., Vafadarnikjoo, A., Reid, I., & Bamford, D. (2025). Unpacking critical success factors to improve supply chain effectiveness, efficiency and performance: A 7Vs framework for consideration. *Production Planning & Control*, 36(4), 521–535.

Bamford, D., Forrester, P., & Reid, I. (2023). *Essential guide to operations management: Concepts & case notes*, 2nd edition. London & New York: Routledge.

Bamford, D., Reid, I., Forrester, P., Dehe, B., Bamford, J., & Papalexi, M. (2024). An empirical investigation into UK University-Industry collaboration: The development of an impact framework. *The Journal of Technology Transfer*, 49, 1411–1443.

Brown, D. (2011). *An experiential approach to organizational development*. Singapore: Pearson.

Bruyaka, O., Prange, C., Li, H.A., (2024). Dynamic capabilities measurement safari: A roadmap. *The International Journal of Management Reviews*, 26, 495–517.

Budler, M., Quiroga, B., & Trkman, P. (2023). A review of supply chain transparency research: Antecedents, technologies, types, and outcomes. *Journal of Business Logistics*, 45.

Celuch, K., Robinson, N., & Walsh, A. (2015). A framework for encouraging retail customer feedback. *Journal of Services Marketing*, 29(4), 280–292.

Chen, J. (2022). Responsible sourcing and supply chain traceability. *International Journal of Production Economics*, 248, 108462.

de Haan, E., Verhoef, P., & Wiesel, T. (2015). The predictive ability of different customer feedback metrics for retention. *International Journal of Research in Marketing*, 32(2), 195–206.

Dehe, B., Bamford, D., & Kotcharin, S. (2022). Bespoke benchmarking framework employed as vehicle and platform for open innovation–a healthcare infrastructure case. *Production Planning & Control*, 35(7), 725–743.

Dienst, S., Fathi, M., Abramovici, M., & Lindner, A. (2014). Development of a knowledge-based feedback assistance system of product use information for product improvement. *IJPD*, 19(4), 191.

Gimenez-Espin, J., Jiménez-Jiménez, D., & Martínez-Costa, M. (2013). Organizational culture for total quality management. *Total Quality Management & Business Excellence*, 24(5–6), 678–692.

Green, T. (2012). TQM and organisational culture: How do they link?. *Total Quality Management & Business Excellence*, 23(2), 141–157.

Gremyr, I., Birch-Jensen, A., Kumar, M., & Löfberg, N. (2022). Quality functions' use of customer feedback as activation triggers for absorptive capacity and value co-creation. *International Journal of Operations & Production Management*, 42, 218–242.

Humeisat, E. (2022). Effect of strategic management practices on organizational excellence. *Journal of Hunan University Natural Sciences*, 49.

Kassela, K., Papalexi, M., & Bamford, D. (2017) 'Applying quality function deployment to social housing?. *The TQM Journal*, 29(3), 422–436.

Maroofi, F. (2012). Link between organisational culture and TQM practices. *International Journal of Intelligent Enterprise*, 1(3/4), 327.

Mirzaei, M., Dehe, B., Weerasinghe, K., & Srivastava, H. (2025). Enhancing supply chain visibility in the fishing industry: The role of emerging technologies and their affordances. *Operations and Supply Chain Management*, 18(2), 146–160.

Mohammad, M., Mann, R., Grigg, N., & Wagner, J. (2011). Business excellence model: An overarching framework for managing and aligning multiple organisational improvement initiatives. *Total Quality Management & Business Excellence*, 22(11), 1213–1236.

Mong, S., Mohamed, S., Misnan, M., & Palis, P. (2021). Integrating resource-based view and performance improvement theory in developing maintenance management continuous improvement model: A conceptual framework. *Studies of Applied Economics*, 39.

Okeke, N., Alabi, O., Igwe, A., Ofodile, O., & Ewim, C. (2024). Customer-centric quality management: A framework for organizational excellence in SMEs. *International Journal of Management & Entrepreneurship Research*, 6(10):3517–3540.

Palumbo, R., & Douglas, A. (2023). The secret ingredient? Uncovering the effect of organizational culture on quality management: A literature review. *International Journal of Quality & Reliability Management*, 41, 195–268.

Papalexi, M., Bamford, D., & Breen, L. (2020) Key sources of operational inefficiency in the Pharmaceutical Supply Chain. *Supply Chain Management: An International Journal*, 25(6), 617–635.

Papalexi, M., Bamford, D., Nikitas, A., Breen, L., & Tipi, N. (2022) Pharmaceutical supply chains and management innovation?. *Supply Chain Management*, 27(4), 485–508.

Parast, M., & Adams, S. (2012). Corporate social responsibility, benchmarking, and organizational performance in the petroleum industry: A quality management perspective. *International Journal of Production Economics*, 139(2), 447–458.

Teece, D.J., (2007). Explicating dynamic capabilities: The nature and microfoundations of (sustainable) enterprise performance. *The Strategic Management Journal*, 28, 1319–1350.

Valmohammadi, C., & Roshanzamir, S. (2015). The guidelines of improvement: Relations among organizational culture, TQM and performance. *International Journal of Production Economics*, 164, 167–178.

3 Managing Quality in the Digital Era

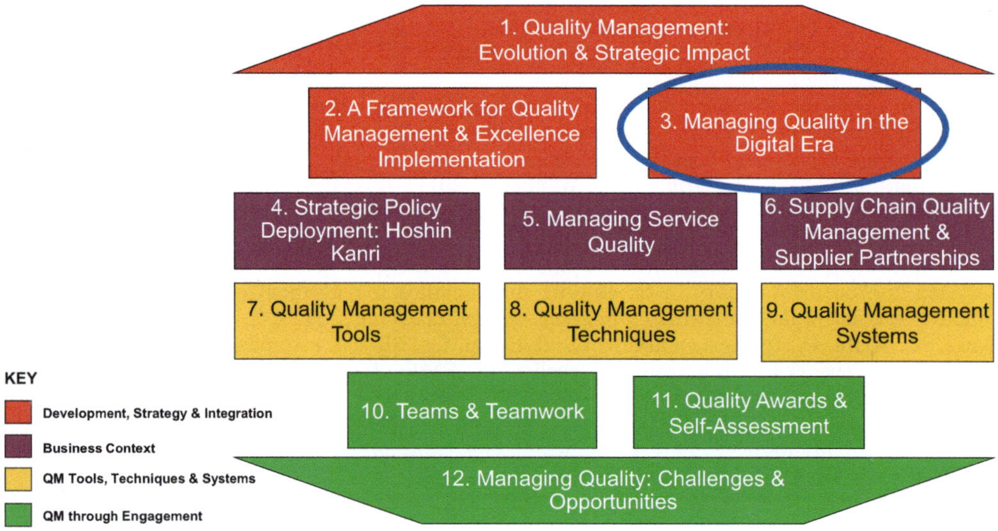

Introduction

As we move into what we can refer to as the post-digital or digital maturity era—a time when digital technologies are no longer a competitive differentiator but an expectation—it is vital to recognise the impact that rapid technological advancement has had on the way organisations operate (Olaniyan et al., 2024; Mirzaei et al., 2025). The influence of the Fourth and emerging Fifth Industrial Revolutions has reshaped business dynamics across sectors. Technologies such as machine learning, automation, and advanced analytics are now embedded in everyday life, from education and finance to healthcare and beyond.

Operational decision-making is increasingly driven by data, necessitating robust digital infrastructures that support the seamless flow of information, knowledge, and collaboration across organisational boundaries (Dehe et al., 2025). In this context, Industry 4.0—characterised by the integration of smart technologies, real-time data exchange, and automation—has enabled organisations to enhance process performance, optimise resource utilisation, and pursue competitive advantage through intelligent systems such as IoT and AI (Cranmer et al., 2022).

DOI: 10.4324/9781003594369-4

As this technological landscape matures, Industry 5.0 is emerging with a renewed emphasis on human-centricity, sustainability, and resilience. It advocates for meaningful collaboration between people and intelligent machines, aiming to personalise experiences and create sustainable value.

Navigating this complex and fast-evolving environment demands more than just technological adoption—it requires a strategic commitment to quality management and organisational excellence. In the digital era, quality is no longer confined to compliance or control; it is about leveraging data intelligently to drive continuous improvement, informed decision-making, and value creation across the enterprise. Organisations that embed quality into their digital transformation journeys will be better positioned to thrive in a world where agility, insight, and resilience are paramount.

Data Quality Management

The collection and analysis of organisational data plays a pivotal role in driving performance, supporting quality management (QM), and enabling continuous improvement (CI). Effective data management not only facilitates the achievement of strategic objectives but also empowers organisations to identify areas for operational enhancement and innovation, thereby strengthening their competitive position (Sebastian-Coleman, 2022).

While the value of data is widely acknowledged across industries (Mohammed et al., 2017), there remains a significant gap in how organisations assess and manage data quality. For instance, Gartner's 2018 Data Quality Tools Magic Quadrant Survey revealed that 57% of respondents did not measure the cost implications of poor-quality data (Chien, 2019). This oversight is not uncommon; as Berman (2008) observed, many organisations lack the capability to manage data quality effectively—particularly in ensuring that the right data is available at the right time and in the right place to support informed decision-making.

The underutilisation of data's full potential is further evidenced by a 2015 report from PricewaterhouseCoopers (PwC, 2015), which highlighted the challenges organisations face in extracting actionable insights from the vast volumes of data they generate and store. These challenges are compounded by the complexity of managing data not only within individual organisations but also across broader business networks. Monitoring, controlling, interpreting, and analysing this data effectively is essential for achieving high performance and sustaining QM and CI initiatives (Redman & Grieves, 1999; Evans & Price, 2018; Aljumah et al., 2021).

To address these challenges, international standards such as ISO 8000–61 have been developed to provide a structured framework for data QM. This standard outlines best practices across key dimensions, including implementation, data-related support, and resource provision (Mohammed et al., 2017). These principles serve as a foundation for organisations seeking to embed quality into their digital infrastructure and ensure that data becomes a strategic asset in the pursuit of organisational excellence. Specifically, these include: the Implementation, Data-Related Support and the Resource Provision (see Figure 3.1).

The first core component of data QM is implementation, as illustrated in Figure 3.1. This is grounded in the well-established Plan-Do-Check-Act (PDCA) cycle, originally developed by Walter A. Shewhart. The PDCA framework provides a structured, iterative approach to improving data quality within a CI context:

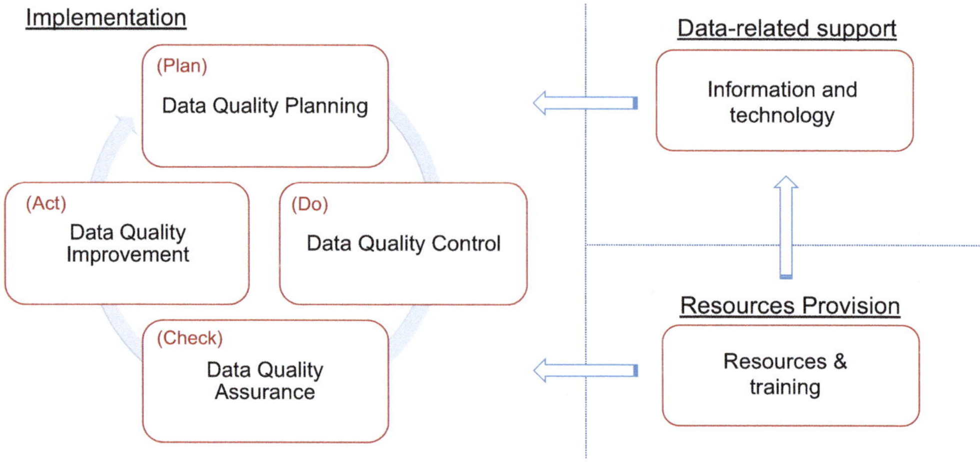

Figure 3.1 Data quality structure.

- Data Quality Planning: At this foundational stage, organisations define a strategic approach to data quality, establishing a robust blueprint that aligns with broader business objectives.
- Data Quality Control: This phase focuses on enhancing the accuracy, consistency, and reliability of data. Key activities include the development of clear data specifications, detailed work instructions, standardised data processing protocols, and the implementation of rigorous monitoring and control mechanisms.
- Data Quality Assurance: Here, the emphasis shifts to evaluating existing processes and identifying opportunities for improvement. This stage supports the refinement of data management practices through performance assessment and process optimisation.
- Data Quality Improvement: The final stage targets the root causes of data quality issues. It involves conducting root cause analysis to prevent future inconsistencies and applying comprehensive data cleansing techniques to rectify existing errors.

The second component of data QM is data-related support, which underpins effective implementation. This includes both technological and procedural enablers across the organisation, such as data architecture, data transfer mechanisms, operational workflows, and data security protocols.

The third component concerns the resources required to manage data quality effectively. This encompasses the allocation of skilled personnel, appropriate tools, and sufficient investment to sustain high standards of data governance.

As Sebastian-Coleman (2022) highlights, effective data QM mitigates the financial and operational risks associated with poor-quality data, while unlocking the potential of well-curated information in today's data-intensive, digital environment. Digitalisation has significantly enhanced the principles underpinning data QM, enabling more agile and responsive data governance. However, it has also introduced new challenges—particularly in relation to data security and integrity (Bakhtari et al., 2021; Hane Hagström et al., 2023).

Data-Driven QM and CI

In today's digital era, it is evident that while most organisations are data-dependent, they are not necessarily data-driven (NewVantage Partners, 2020). Over recent years, considerable effort has been directed toward achieving data-driven transformation, as documented by the International Data Group (2018, 2020). This transformation involves the development and integration of digital capabilities that enable organisations to adopt emerging technologies, reconfigure operational systems, enhance improvement processes, and ultimately create greater value (Edu et al., 2020; Feitosa Jorge et al., 2024; Mitchell et al., 2025).

However, the journey toward becoming truly data-driven is not without its challenges. Sebastian-Coleman (2022) identifies five key barriers that organisations must address to realise the full potential of data-driven transformation. These challenges are presented in Figure 3.2 and serve as a useful framework for understanding the strategic and operational complexities involved in embedding data-centric thinking across the enterprise.

Organisations are currently inundated with vast volumes of data. While this abundance presents opportunities for insight and innovation, it also introduces significant challenges in terms of management and value creation (Dehe et al., 2025). As Dodds (2020) and Press (2016) observed, a considerable portion of data scientists' time is consumed by locating and preparing the data they require—an inefficiency that reflects deeper systemic issues in data

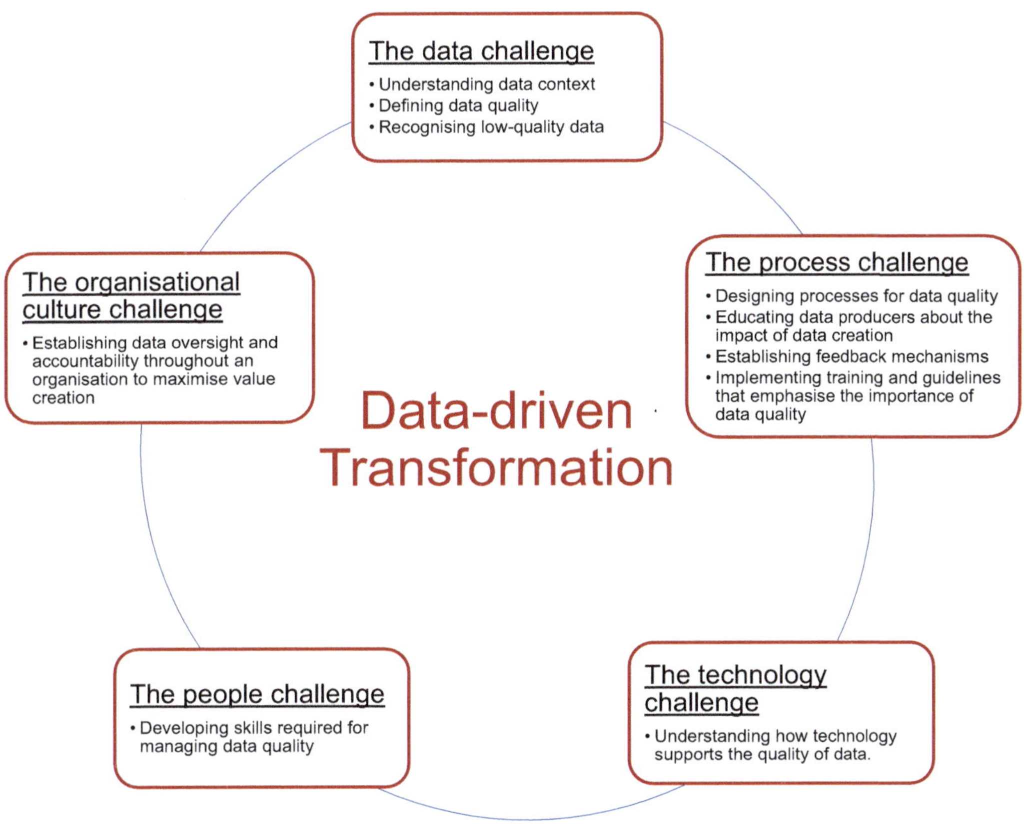

Figure 3.2 Digital transformation key challenges.

governance. This is understandable, given the inherent risks associated with poor data management, which can compromise operational integrity and strategic decision-making (Rosak-Szyrocka et al., 2024).

Poor-quality data is more than a technical inconvenience; it is a strategic liability. It can lead to delays, rework, misinformed decisions, and ultimately, a decline in organisational reputation and customer satisfaction. These outcomes directly undermine the principles of QM, which emphasise consistency, reliability, and CI in all aspects of organisational performance.

To mitigate these risks and elevate data from a passive resource to a strategic asset, organisations must adopt a structured approach rooted in organisational excellence. This involves embedding quality principles into data governance frameworks, ensuring that data is accurate, timely, and fit for purpose. Moreover, it requires a cultural shift—where data is not only managed but actively leveraged to drive innovation, agility, and competitive differentiation.

One pathway to achieving this is through digitalisation, which offers tools and technologies capable of transforming data management practices. However, one of the central challenges in this transformation is the selection of appropriate, innovative solutions that align with organisational goals and capabilities. When approached through the lens of QM, digital tools can be evaluated not just for their technical features but also for their ability to support CI, stakeholder value, and long-term resilience.

The emergence of Industry 4.0 technologies has significantly influenced the evolution of QM, giving rise to what is now referred to as 'Quality 4.0'. These technologies—such as Artificial Intelligence (AI), Machine Learning (ML), the Internet of Things (IoT), and Virtual Reality (VR)—enable organisations to collect, store, analyse, and share data more effectively, while also addressing critical challenges such as data security and integrity (Cranmer et al., 2022; Sebastian-Coleman, 2022). By facilitating the creation of new data types and perspectives, these tools support more informed decision-making and foster collaboration across increasingly complex supply networks.

The integration of Industry 4.0 technologies into operational environments allows for more efficient, flexible, and sustainable production systems, which are increasingly tailored to individual customer needs (Rosak-Szyrocka et al., 2024; Bibby & Dehe, 2018). This shift marks a departure from traditional QM approaches, as organisations begin to embed digital capabilities into their quality systems, aligning with international standards such as ISO and advancing broader strategic goals around transparency, resilience, and sustainability (Chiarini, 2020; Javaid et al., 2020).

Quality 4.0 represents a transformation in how quality is understood and applied. It moves beyond compliance and control, positioning quality as a strategic enabler of organisational excellence. Through the automation and integration of digital operations, businesses can enhance performance across the value chain. For example, in the footwear industry, the implementation of a digital supply chain has demonstrated tangible benefits:

- Improved collaboration and data sharing across the supply network through integrated digital systems (e.g., Product Lifecycle Management platforms).
- Increased transparency via access to previously unavailable data, including sustainability metrics, ethical sourcing practices, and climate impact assessments.
- Reduced lead times—from 60 to 10 days—by using VR in prototyping, eliminating the need for multiple physical samples and reducing material waste.

- Enhanced customer satisfaction through customisation and ethical transparency, enabled by VR-based platforms that allow customers to engage directly with the production process.

Ultimately, Quality 4.0 strengthens collaboration across supply networks, improves productivity control, and reduces operational risk. By embedding Industry 4.0 technologies into QM systems, organisations are better equipped to achieve resilience, agility, and organisational excellence in an increasingly digital and competitive global environment.

The integration of Industry 4.0 technologies has made a substantial contribution to the structure and implementation of data QM, as outlined in Figure 3.3, which builds upon the foundational framework presented in Figure 3.1. These technologies accelerate the implementation phase by enhancing the precision, speed, and reliability of data collection, analysis, and communication. In doing so, they support the development of smart factories—digitally enabled environments that optimise various dimensions of production, operations, and supply chain management (Alexa et al., 2022).

From a QM perspective, Industry 4.0 tools enable organisations to embed data integrity and process control into their operational systems, thereby reducing variability and improving consistency. This supports the broader pursuit of organisational excellence, where data becomes a strategic asset used to drive CI, informed decision-making, and stakeholder value.

However, the adoption of these technologies also introduces new complexities. As Rosak-Szyrocka et al. (2024) note, the benefits of digitalisation must be balanced against the need to adapt these innovations to a wider socio-economic context. This includes considerations around workforce capability, ethical data use, and the alignment of digital systems with organisational culture and strategic objectives. So, while Industry 4.0 technologies offer powerful tools for enhancing data quality and operational performance, their successful implementation requires a holistic approach—one that integrates digital capabilities with the principles of QM and the strategic imperatives of organisational excellence.

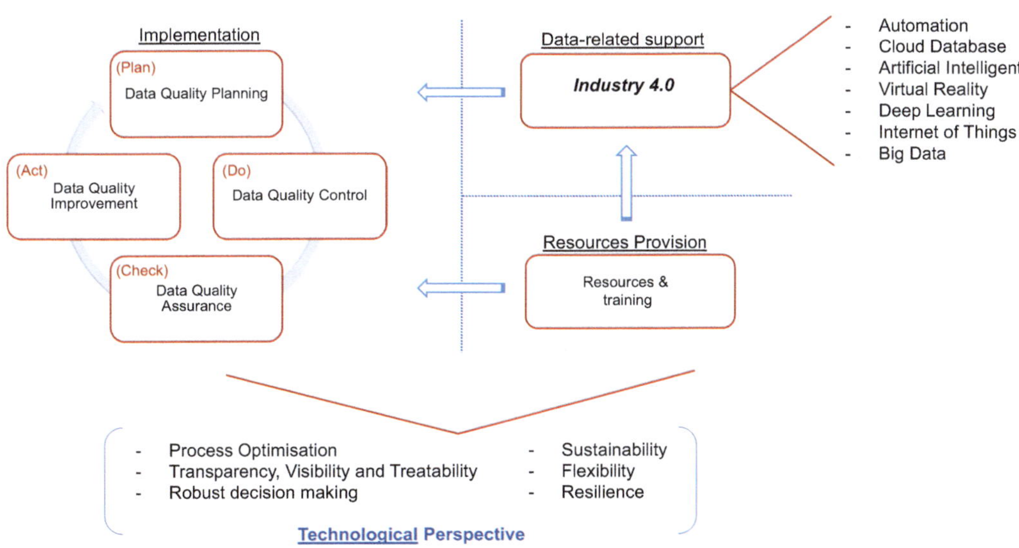

Figure 3.3 Industry 4.0—data quality structure.

The evolving emphasis on sustainability, resilience, and human-centric innovation has led the European Commission to propose Industry 5.0 as a natural progression from Industry 4.0. This next phase prioritises environmentally responsible technologies, inclusive social practices, and robust organisational mechanisms (European Commission, 2021; Banholzer, 2022). Industry 5.0 has already catalysed transformation across multiple sectors—including textiles, education, and healthcare—particularly within developed economies. It has advanced digital supply chains, enhanced corporate infrastructures, improved network communication, and promoted transparency, all of which contribute meaningfully to the pursuit of quality and organisational excellence.

Importantly, Industry 5.0 supports the development of sustainable strategies that address diversity, fair trade, equality, well-being, access to education, and the reduction of poverty and waste. These values align closely with the broader goals of QM, which extend beyond operational efficiency to encompass stakeholder satisfaction, ethical governance, and long-term value creation. However, the implementation of Industry 5.0 principles varies significantly across global regions. Advanced economies—particularly in Europe and the Western hemisphere—have demonstrated greater capacity to adopt these innovations, largely due to their existing socio-economic infrastructure. In contrast, developing nations such as India face considerable barriers, including socio-economic inequality and underdeveloped physical infrastructure (Rosak-Szyrocka et al., 2024).

The overarching aim of Industry 5.0 is to evolve the smart factory—established under Industry 4.0—into a more sustainable and human-centric ecosystem. This transformation is guided by three strategic priorities:

- Circular Economy: Promoting environmental stewardship through resource recycling and renewable energy. For example, IoT devices powered by sustainable energy sources reduce environmental impact, while AI supports emissions reduction by identifying optimal operational practices (Varjovi & Babaie, 2020; Fraga-Lamas et al., 2021).
- Society: Creating workplaces that prioritise creativity, responsibility, and well-being. Human–machine collaboration, such as the use of robotics to reduce biomechanical risk, exemplifies this shift (Tiseni et al., 2021).
- Flexibility and Resilience: Enhancing customer satisfaction through personalisation and customisation. Technologies such as 3D printing and AI have revolutionised healthcare by enabling the production of personalised smart implants, improving surgical outcomes (Adel, 2022).

Despite these advancements, the ongoing transformation raises critical concerns—particularly around cybersecurity. As businesses become increasingly digital, the risk of cyberattacks grows. The proliferation of remote work, cloud computing, and IoT devices has expanded the attack surface, making cybersecurity a pressing challenge (Fatima et al., 2022; Hane Hagström et al., 2023). According to Statista, the average cost of the most disruptive cybersecurity breach in UK businesses in 2024 was £1,205, though the true cost is often far greater when reputational damage and recovery time are considered. High-profile incidents, such as Apple's $95 million settlement over privacy violations and Krispy Kreme's system disruption due to a cyberattack, underscore the urgency of this issue.

To address these risks, both Industry 4.0 and 5.0 place increasing emphasis on designing augmented intelligence security architectures that protect end users and ensure data

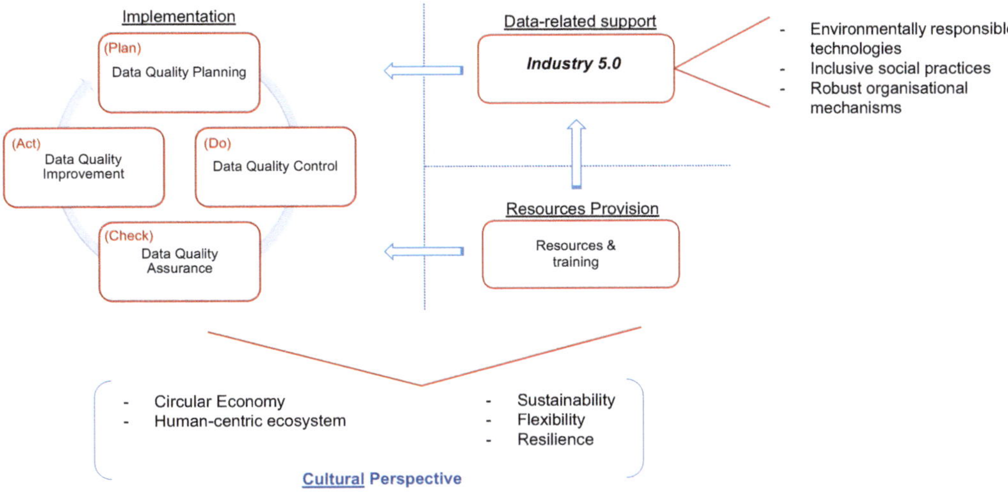

Figure 3.4 Industry 5.0—Data quality structure.

integrity (Chergui et al., 2021). However, the challenges extend beyond cybersecurity. The increased computational demands of digital technologies contribute to higher energy consumption (Hazra et al., 2021), and their social implications—such as workplace isolation, skill redundancy, and diminished employee engagement—must also be carefully managed (Doyle Kent and Kopacek, 2021).

As illustrated in Figure 3.4, Industry 5.0 introduces a more environmentally conscious, socially inclusive, and human-centric approach to organisational transformation. It seeks to harmonise efficiency and productivity with ecological preservation and social responsibility, thereby fostering sustainable and resilient ecosystems. For organisations committed to QM and excellence, this represents not just a technological shift, but a cultural and strategic imperative.

Summary

This chapter has explored the profound transformation of business operations in the context of globalisation, digitalisation, and the transition into the post-digital era. It has highlighted how the convergence of advanced technologies—particularly those associated with Industry 4.0 and Industry 5.0—has reshaped organisational structures, supply chains, and decision-making processes.

Central to this transformation is the emergence of data-driven QM, which enables organisations to orchestrate complex, interconnected systems with greater precision and agility. Technologies such as the IoT, AI, and ML have evolved beyond operational tools; they now serve as strategic enablers of organisational excellence, driving automation, connectivity, and real-time responsiveness.

The concept of Quality 4.0 reflects this shift, integrating digital capabilities into traditional quality frameworks to enhance transparency, flexibility, and resilience. Meanwhile, Industry 5.0 introduces a more human-centric and sustainable approach, aligning technological innovation with social responsibility and ecological stewardship.

These developments support the creation of smart, adaptive ecosystems that prioritise stakeholder value and long-term sustainability.

However, the chapter also acknowledges the c hallenges that accompany digital transformation—particularly in relation to cybersecurity, energy consumption, and the social implications of automation. These risks underscore the importance of embedding QM principles into digital strategies, ensuring that technological adoption is guided by ethical governance, CI, and a commitment to excellence.

Looking ahead, organisations that succeed in the post-digital phase will be those that treat data not merely as a resource, but as a strategic asset. By aligning digital innovation with robust quality systems and a culture of excellence, businesses can build resilient, adaptive, and purpose-driven operations capable of thriving in an increasingly complex global environment.

References

Adel, A. (2022). Future of Industry 5.0 in society: Human-centric solutions, challenges and prospective research areas. *Journal of Cloud Computing: Advances, Systems and Applications*, 11, 40.

Alexa, L., Pîslaru, M., & Avasilcăi, S. (2022). From Industry 4.0 to Industry 5.0: An overview of European Union enterprises. In *Sustainability and Innovation in Manufacturing Enterprises* (pp. 221–231). Springer.

Aljumah, A., Nuseir, M., & Alam, M. (2021). Organizational performance and capabilities to analyze big data: Do the ambidexterity and business value of big data analytics matter?. *Business Process Management Journal (BPMJ)*, 27, 1088–1107.

Bakhtari, A.R., Waris, M.M., Sanin, C., & Szczerbicki, E. (2021). Evaluating Industry 4.0 implementation challenges using interpretive structural modeling and fuzzy analytic hierarchy process. *Cybernetics and Systems*, 52(5), 350–378.

Banholzer, V.M. (2022). From Industry 4.0 to Society 5.0 and Industry 5.0: Value- and Mission-Oriented Policies: Technological and Social Innovations-Aspects of Systemic Transformation (IKOM WP Vol. 3, No. 2/2022). Nurnberg: Technische Hochschule Nurn-berg Georg Simon Ohm.

Berman, F. (2008). Got data? A guide to data preservation in the information age. *Communications of the ACM*, 51(12), 50–56.

Bibby, L., & Dehe, B. (2018). Defining and assessing industry 4.0 maturity levels–case of the defence sector. *Production Planning & Control*, 29(12), 1030–1043.

Chergui, H., Blanco, L., Garrido, L.A., Ramantas, K., Kukliński, S., Ksentini, A., & Verikoukis, C. (2021). Zero-touch AI-driven distributed management for energy-efficient 6G massive network slicing. *IEEE Network*, 35(6), 43–49.

Chiarini, A. (2020). Industry 4.0, quality management and TQM world: A systematic literature review and a proposed agenda for further research. *The TQM Journal*, 32(4), 603–616.

Chien, M. (2019). Drive Data Quality Improvement from a Foundation of Metrics. Gartner Research Note. ID G00378494. (PDF). July 22, 2019.

Cranmer, E.E., Papalexi, M., Tom Dieck, M.C., & Bamford, D., 2022. Internet of Things: Aspiration, implementation and contribution. *Journal of Business Research*, 139, pp.69–80.

Dehe, B., Subasinghage, M., & Mirzaei, M. (2025). Orchestrating resources for Big data analytics implementation in manufacturing SMEs: insights into managerial role and engagement. Production Planning & Control, 1–33.

Dodds, L. (2020). Do Data Scientists Spend 80% of Their Time Cleaning Data? Turns Out, No? (Lost Boy: The Blog of @ldodds). January 31, 2020. https://blog.ldodds.com/2020/01/31/do-data-scientists-spend-80-of-their-time-cleaning-data-turns-out-no/#:~:text=variant%20of%20it%3A-,Data%20scientists%20spend%2080%25%20of%20their%20time,data%20rather%20than%20creating%20insights.&text=Data%20scientists%20only%20spend%2020,data%20quality%2C%20standards%2C%20access

Doyle Kent, M., & Kopacek, P. (2021). Do we need synchronization of the human and robotics to make Industry 5.0 a success story? In *The International Symposium for Production Research* (pp. 302–311). Cham: Springer.

Edu, S.A., Agoyi, M., & Agozie, D.Q. (2020). Integrating digital innovation capabilities towards value creation. *International Journal of Intelligent Information Technologies*, 16(4), 37–50.

European Commission (2021). *Industry 5.0 Towards a Sustainable, Human-centric and Resilient European Industry*. Brussels: European Union.

Evans, N., & Price, J. (2018). Death by a thousand cuts: Behaviour and attitudes that inhibit enterprise information asset management. *Information Research*, 23, paper 779.

Fatima, Z., Tanveer, M.H., Zardari, S., Naz, L.F., Khadim, H., Ahmed, N., & Tahir, M., 2022. Production plant and warehouse automation with IoT and Industry 5.0. *Applied Sciences*, 12(4), 2053.

Feitosa Jorge, L., Mosconi, E., & Santa-Eulalia, L.A.D. (2024). Enterprise social media to foster digital maturity: A value-creation perspective. *Journal of Systems and Information Technology*, 26(3), 313–336.

Fraga-Lamas, P., Lopes, S.I., & Fernández-Caramés, T.M., (2021). Green IoT and edge AI as key technological enablers for a sustainable digital transition towards a smart circular economy: An Industry 5.0 use case. *Sensors*, 21(17), 5745.

Hane Hagström, M., Bergsjö, D., & Wahrén, H. (2023). Barriers from a socio-technical perspective to implement digitalisation in industrial engineering processes: Literature review. *Proceedings of the Design Society*, 3, 737–746.

Hazra, A., Adhikari, M., Amgoth, T., & Srirama, S.N., 2021. Collaborative AI-enabled intelligent partial service provisioning in green industrial fog networks. *IEEE Internet of Things Journal*, 10(4).

International Data Group. (2018). The State of Digital Business Transformation, 2018. (White Paper). April 22, 2018. https://www.idg.com/news/the-state-of-digital-business-transformation-2018/

International Data Group. (2020). 2020 State of the CIO. January 23, 2020. https://www.idg.com/tools-for-marketers/2020-state-of-the-cio/

Javaid, M., Haleem, A., Vaishya, R., Bahl, S., Suman, R., & Vaish, A. (2020). Industry 4.0 technologies and their applications in fighting COVID-19 pandemic. *Diabetes & Metabolic Syndrome: Clinical Research & Reviews*, 14(4), 419–422.

Mirzaei, M., Dehe, B., Weerasinghe, K., & Srivastava, H. (2025). Enhancing supply chain visibility in the fishing industry: The role of emerging technologies and their affordances. *Operations and Supply Chain Management*, 18(2), 146–160.

Mitchell, T., Bamford, D., Papalexi, M., Reid, I., & Webster, R. (2025). Impact pathways: High-performance sports operationalisation through Industry 4.0: An actionable blueprint. *International Journal of Operations and Production Management*, *45*(13), 253–265.

Mohammed, A.G., Eram, A., & Talburt, J.R. (2017). ISO 8000-61 data quality management standard, TDQM compliance, IQ principles. In *MIT International Conference on Information Quality*. Little Rock, USA: University of Arkansas.

NewVantage Partners. (2020). Big data and AI executive survey 2020: Executive Summary of Findings. Foreword by Thomas Davenport and Randy Bean. https://www.newvantage.com/thoughtleadership

Olaniyan, O., Dehe, B., Bamford, D., & Ward, S. (2024). Enhancing on-pitch learning capabilities with data analytics and technologies in elite sports. *European Sport Management Quarterly*, *24*(6), 1195–1214.

Press, G. (2016). Cleaning big data: Most time-consuming, least enjoyable data science task, survey says. *Forbes*. March 23, 2016. https://www.forbes.com/sites/gilpress/2016/03/23/data-preparation-most-time-consuming-least-enjoyable-data-science-task-survey-says/#30a7c57a6f63

PriceWaterhouseCoopers (PwC) (2015). Seizing the Information Advantage—Executive Summary. How Organizations Can Unlock Value and Insight from the Information They Hold. (White Paper in conjunction with Iron Mountain). https://www.ironmountain.com/resources/whitepapers/s/seizing-the-information-advantage-executive-summary.aspx

Redman, T., & Grieves, J. (1999). Managing strategic change through TQM: Learning from failure. *New Technology, Work and Employment*, 14(1), 45–61.

Rosak-Szyrocka, J., Żywiołek, J., & Shahbaz, M. (2024). *Quality Management, Value Creation, and the Digital Economy*. Abingdon, Oxon, New York, NY: Routledge.

Sebastian-Coleman, L. (2022). *Meeting the Challenges of Data Quality Management*. London: Academic Press.

Tiseni, L., Chiaradia, D., Gabardi, M., Solazzi, M., Leonardis, D., & Frisoli, A., (2021). UV-C mobile robots with optimized path planning: Algorithm design and on-field measurements to improve surface disinfection against SARS-CoV-2. *IEEE Robotics & Automation Magazine*, 28(1), 59–70.

Varjovi, A.E., & Babaie, S. (2020). Green Internet of Things (GIoT): Vision, applications and research challenges. *Sustainable Computing: Informatics and Systems*, 28, 100448.

Part 2

The Business Context of Quality Management

4 Strategic Policy Deployment

Hoshin Kanri

Introduction

An increasing number of organisations, as part of a strategic planning approach, are using policy deployment, often operationalised through 'Hoshin Kanri' (HK), which serves as a strategic quality management framework. The process enables being consistent in the alignment of target-setting and achievement, not just in magnitude but in the overall organisational effectiveness.

In the late 1980s, the concept of policy deployment was little known outside Japan. Dale (2016) recalled leading study missions of European executives and management consultants in Japanese manufacturing companies in 1988 and 1989, and it was clear that, when the concept of 'HK' or policy deployment was introduced by the host organisation, this was something new to the participants. Today, while the concept of 'HK' is more common, many organisations do not fully embrace it (Wilson et al., 2024).

Policy deployment is an English translation of 'HK', the Japanese strategic planning and management process involving setting direction and deploying the means of achieving

DOI: 10.4324/9781003594369-6

that direction. The PDCA (Plan-Do-Check-Act) cycle, often associated with lean thinking and continuous improvement logic, is used extensively in the process.

In the 2000s, HK started to be more established in the USA and Europe, after waves of Deming Application Prize-winning Japanese subsidiaries such as Hewlett-Packard's YHP Division and Fuji-Xerox and the implementation of lean management and six-sigma programmes. Today, the Malcolm Baldrige National Quality Award (MBNQA), the European Foundation for Quality Management Global Award and the Shingo Prize (c.f. Chapter 11), which are considered the most prestigious 'Excellence Awards', strongly promote the concepts associated with strategy policy deployment. Award-winning companies (e.g. Freese and Nichols, Inc., Vamed, or Ipsen) consistently employ policy deployment and HK successfully to link medium- to long-term strategies to annual plans to achieve significant improvements in business results. According to Santhiapillai and Ratnayake (2023), policy deployment enables creating systems and cohesion within a business that is transparent and consistent throughout the company. It provides a platform to identify and align strategic goals and objectives across the different departments and teams and all levels (Mothersell et al., 2008). Finally, the X-Matrix is a visual tool used within HK to clearly display the relationships between strategic objectives, key performance indicators, action items, and ownership, essentially acting as a single-page roadmap for achieving strategic goals, which we will detail in this chapter.

This chapter reviews the definitions, concepts, and characteristics of policy deployment and proposes a policy deployment model based on the catch, reflect, improve, pass (CRIP) process, adapted from Mulligan et al. (1996), evidencing the catchball technique (Paraschivescu & Stoica, 2018). The catch, reflect, improve, scrutinise, and then pass (CRISP) model, based on research described in Lee and Dale (1999), demonstrates how policy deployment, business results and self-assessment are inextricably linked into the PDCA cycle, as per Figure 4.1, and it is argued that this will enable an organisation to deploy its vision, mission, goals, objectives, targets and means more effectively.

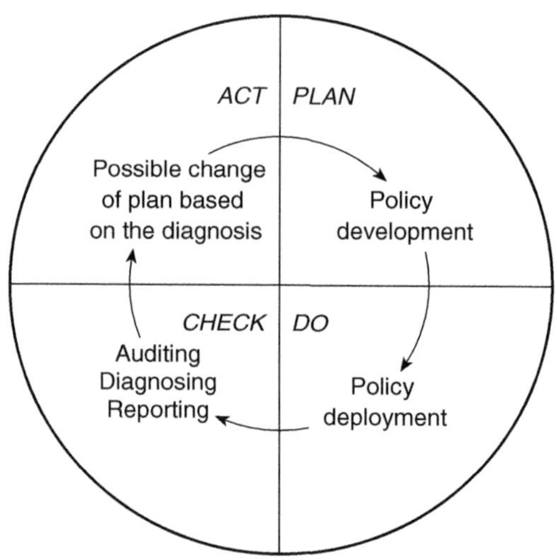

Figure 4.1 The plan-do-check-act cycle. Source: Dale et al. (2016).

Definitions of Policy Deployment or Hoshin Kanri (HK)

Policy deployment or HK was developed to communicate a company's strategy, goals and objectives throughout its hierarchy. Its main benefit is to focus attention on prioritising and aligning key activities for success (Mothersell et al., 2008; Marksberry, 2011). A literal translation of HK provides an insight into its concept (Jolayemi, 2008):

Hoshin = a direction, a compass, a plan, an aim. It represents the strategic direction or guiding principle.
Kanri = management or control. It refers to the act of managing or controlling.
Together, HK means 'management and control of the company's focus'.

The term 'policy deployment' is often used as the translation of, and/or used interchangeably with 'HK'. While the translation is useful as an insight, particular attention should be drawn to the 'management', 'control,' and 'focus' elements. While there are several varying definitions, it can be agreed that HK is a structured approach, process or system to strategic planning that aligns an organisation's vision, priorities, and daily operations. It ensures that breakthrough goals are effectively deployed, integrated into daily management, and continuously reviewed for progress. While senior management sets strategic direction, middle and frontline managers play a crucial role in driving its deployment and continuous improvement (Wood & Munshi, 1991; Butterworth & Witcher, 2001; Wilson et al., 2024).

The policy deployment system relies on a series of concepts and techniques (e.g. PDCA cycle, X-Matrix, catchball, and feedback loops) to provide a one-year plan for achieving objectives developed in conjunction with management's choice of specific targets and means in quality, cost, delivery, innovation, resilience, sustainability and morale. This system enables the organisation to conduct work according to the plan, and evaluate, investigate and provide feedback on the results to continuously improve its performance.

The following are the main characteristics which emerge from policy deployment definitions:

- Clear organisational goals are understood by employees at each level of the organisation.
- Involvement of employees in the development of action plans.
- All employees and functions work in the same direction to achieve the organisation's objectives.
- Regular review mechanism using the PDCA cycle and promoting an environment of continuous improvement.
- Visible display of information.
- Feedback loop and consensus-based development.

How Does Policy Deployment Work?

Policy deployment within a process of long-term planning is one of the features of world-class organisations that exhibit operational excellence capabilities. The process involves developing plans, targets, controls and areas for improvement based on the previous level's policy and an assessment of the previous year's performance. The plans and targets are discussed and debated at each level until a consensus on plans and targets is reached (process also referred to as catchball), along with the methods for meeting the goal. Jolayemi (2008) provides the following definition of catchball.

> Catchball refers to relative up, down, and horizontal discussions and joint analysis necessary for effective determinations of objectives, strategies, targets and means.

The control of the deployment process and subsequent implementation of the policies is conducted through quarterly, monthly, weekly and daily reviews depending on the level of the individual involved. The X-Matrix is an associated visual tool that enables the capture and summarisation of this information into one document.

The PDCA cycle is extensively used in these diagnoses and the discipline of policy deployment and agreement at each level, ensuring everyone is working in the same direction (Dale, 1990; Butterworth & Witcher, 2001). Kondo (1998) notes that HK proved extremely effective in furthering company-wide improvement plans by uniting the efforts of all employees.' The most important policy deployment concepts to be drawn from Dale (1990), Kondo (1998), Melander et al., (2016) and Giordani da Silveira et al., (2017) are:

- Leadership
- Communication
- Control
- Review

And that there are four stages:

- Policy setting
- Policy deployment
- Policy implementation
- Evaluation, audit and feedback

However, there are several elements which are key but have the tendency to be overlooked or underplayed by some organisations, for example:

- Policy deployment is an effective tool to motivate and engage employees.
- The process is based on two-way communication as per the catchball principle.
- It allows senior managers to exercise leadership and communicate directly and transparently with all employees.
- Evaluations and audits are used to find and solve problems, discover and build on strengths, and standardise and institutionalise improvements. If evaluations and audits are carried out in the wrong way, they will become superficial and ritualistic.

X-Matrix

The X-Matrix is a key framework used in policy deployment to visually connect strategic goals with operational execution. It serves as a structured and visual roadmap, ensuring alignment between long-term vision, annual objectives, key initiatives, and individual responsibilities. This matrix facilitates a top-down and bottom-up approach to strategic policy deployment, ensuring that all organisational levels contribute to achieving the 'Vital Few Objectives' (VFOs). The X-Matrix is a visual planning tool arranged in an 'X' shape, as illustrated in Figure 4.2, where each quadrant represents a different level of strategic planning. These sections are interconnected, showing how high-level strategy translates into specifications.

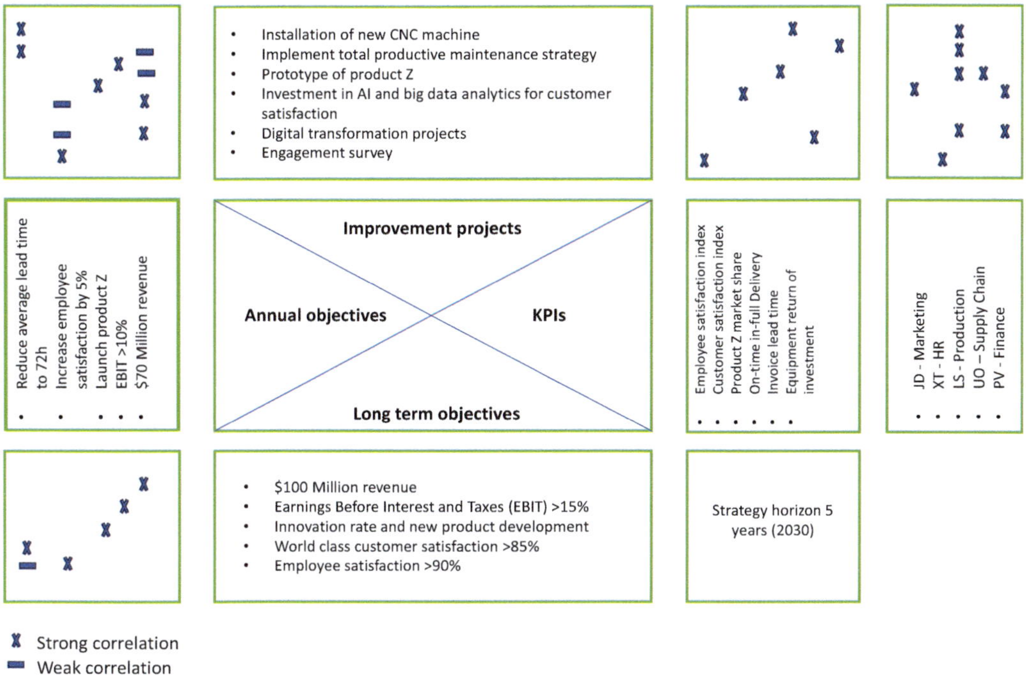

Figure 4.2 Example of an organisation X-Matrix.

The long-term strategic objectives are usually presented in the South Quadrant (bottom). They capture the organisation's 3- to 5-year strategic goals.

- Typically focused on quality improvement, customer satisfaction, operational efficiency, or market expansion.

The annual objectives are presented in the West Quadrant (left).

- They define the specific goals to be achieved within the current year to move toward the long-term strategy.
- These should be SMART (Specific, Measurable, Achievable, Relevant, Time-bound).

The key initiatives or prioritised improvement projects are mapped in the North Quadrant (top).

- They identify the critical projects or actions necessary to achieve the annual objectives.
- They ensure cross-functional collaboration and resource alignment.

The metrics, KPIs and accountability are composed in the East and Far East Quadrants (right).

- They specify the performance indicators used to measure success.
- They assign ownership and responsibility to teams or individuals.

The X-Matrix is useful to promote: i) strategic alignment, as it ensures that all improvement initiatives directly contribute to the organisation's objectives; ii) cross-functional collaboration, as it encourages communication between departments, reducing silos; iii) data-driven decision-making, as it links initiatives to measurable KPIs, ensuring objective performance assessment, and iv) continuous improvement, as it enables the organisation to review progress regularly and adjust strategies accordingly.

The X-Matrix is a powerful tool for integrating strategic planning and quality management. Ensuring a clear connection between long-term vision and daily execution helps organisations maintain focus, improve accountability, and drive continuous quality improvements.

The Policy Deployment Process

Policy deployment works on two levels to manage continuous improvement and achieve business results: strategic objectives and daily control of the business. The key features of the process are now examined.

Five- to Ten-Year Vision

A challenging, customer-focused vision, pertinent to employees at all levels and appropriate for the next five to 10 years, is required. Often a draft of the vision is shared within the organisation for a 'reality check' and feedback, and once validated is then communicated to everyone. This is easier said than done, as visions are generally created at the senior management level and any reality check and/or feedback may receive middle management filtering of employees' comments to prevent unfavourable views reaching top management (i.e. the 'sponge' effect). An effective way of overcoming this problem is to gather accurate information on the company, its customers, competitors and market, and then hold workshops between senior management strategists and all employees.

Mid-term Three- to Five-Year Objectives

Translating the vision into mid-term objectives, together with the broad means to achieve them, is the next step. The objectives should be prioritised, and from this process, the critical ones are selected with a focus on a small number of breakthrough objectives (we would recommend three to five maximum). Then, determine how the objectives will be achieved and cascade the objectives and means through catchball discussions. The process of catchball provides the opportunity to ensure commitment to objectives at each hierarchical level and produces an organisation which is focused and committed to aligned goals. However, these medium-term goals, as extensions of past performances, are of little value without analysis of critical problems, current practices and changes inside and outside the company.

Annual Plan and Objectives

Annual, short-term objectives are determined from the mid-term goals, and the annual plans should be actionable and specific. This one-year plan includes the targets, means and measures that each manager will work on during that year. The managers will develop their annual policies, improvement targets and plans for every section and department which are within their remit of responsibility. It is highly recommended to choose a small number of target areas on which to focus (we would recommend six to eight maximum), and half of these should be related to the manager's participation in the strategic plan and the other half to the critical process of the employee's regular job. Regardless of the number of objectives, all must be measurable, with monthly numerical targets, and the reasons for selection must be compelling and obvious (Jolayemi, 2008). Furthermore, they should be owned by the organisation through the process of catchball, and the plan, objectives, and targets should not be constrained at the management level, but cascaded down to each team or employee.

Deployment/Roll Down to Departments

Clear, disciplined action plans with direction for improvement, what is to be measured, and the processes to be improved are generated through a continuous catchball between all organisational levels and around chosen targets. Corporate and division/department planning cycles should be synchronised, and annual plans should present a prioritised set of actionable tasks, designed to achieve breakthroughs in critical areas (Jolayemi, 2008). A significant aspect of policy deployment is the extent to which the targets and means initiated at the top level are extensively modified through negotiation by the creativity of the lower levels of employees through a bottom-up feedback loop. This involvement of everyone results in full ownership and understanding of the plan at all levels of the organisation. The goal of catchball is to prevent sub-optimisation, and local optimisation may have to be forsaken in the interests of the company, even if this appears to rebuff the concept of 'empowerment' which pervades modern organisations. Once the goals, objectives and plans have been agreed upon, along with the methods and means to reach the goal, they should be openly displayed in the work area adjacent to the progress charts that are tracking achievement and targets and be part of the visual management system.

Execution

The actionable tasks, following the deployment phase, should be taken up by teams and individuals, departmentally and cross-functionally, depending on the task. When using the cross-functional approach, it is essential to determine the responsibility for and supervision of the task. The team/individual should have a clear statement of target and means and follow a PDCA cycle, with periodic checks from senior management (Wood & Munshi 1991; Jolayemi, 2008). It is usual for a department to keep a register of the improvement action agreed with staff. In this way, the improvement activities of an organisation are focused on carrying out projects to meet these policies. This ensures integration and the internalisation of objectives. Every employee understands their manager's policy and therefore knows what to do and that everyone is working in a common and unified direction, or alignment, to achieve the goals of the business. In addition, employees understand the issues which are important to the company, helping to facilitate relations between the different sections of the business.

Progress Review (Monthly and Quarterly)

The control and review of policies are necessary to compare actual to planned performance of each activity that is supporting the organisation's goals and objectives; identify gaps, problems and root causes; determine countermeasures; and recognise and reward achievement.

The progress reviews stress the importance of self-diagnosis of targets and processes to ensure that there is an alignment of activities in the cascading of organisational objectives and operating methods (Mothersell et al., 2008). Problems should be identified and corrective action implemented. PDCA is built into the policy deployment process, and regular checking will assure continuous improvement and reduce costs (see Figure 4.2). The results of these activities are reflected in the following year's policy and assist in improving the process of deployment. In some organisations, the process of policy deployment is also subject to diagnosis by outside experts.

Annual Review

Dale et al., (2016) suggest the review process should focus on the following:

- Achievements of the past year
- Lessons learned in the past year
- The gap between goals and achievements in the past year
- Root cause analysis of the problems
- Environmental factors
- Future plans for the organisation

This forms the basis of policy deployment for the succeeding year, the check being undertaken at the start of the cycle. However, one important review should not be forgotten, the review of the policy deployment process itself, to learn from mistakes made and improve it for the following year; the process can also constitute a useful theme for a benchmarking project (e.g. Dehe et al., 2024).

Dale et al., (2016) say that one of the most tangible aspects of policy deployment is the four sets of reports that support the organisation's planning process:

Hoshin Action Summary

- Articulated objectives
- Objective owners
- Long- and short-term goals
- Implementation strategy
- Specific improvement focus

Hoshin Action Plan

- Detailed links between core objectives and implementation initiatives

Hoshin Implementation Plan

- Records progress as the plan is implemented
- One plan for each objective
- Incorporates: task ownership; milestones and due dates

Hoshin Implementation Review

- Charts post-implementation results relative to company goals
- Competitive benchmarks
- Accepted world-class benchmarks

Visible Display and Visual Management

It is good practice for a department or business unit to have a visible display of the outcomes from the policy deployment process as part of their visual management system. Figure 4.3 illustrates the key points of the typical format of such a display. The left-hand side of the chart shows the tree of policy deployment from the plant manager down to the level of each section. The overall rate of imperfection for each section is related to the different processes, with information being provided on individual problems. A proportion defective (p) control chart is used to monitor the rate of imperfection against the set target. The right-hand side of the chart indicates the annual improvement targets for quality, cost, delivery, safety and morale. A slogan relating to the improvement is displayed at the head of the problem to be solved. The names of the workers who are responsible for the various activities relating to the policy deployment are also displayed. This repository logs each improvement which has been made and helps to promote standardisation. It also serves as a point of reference as to what type of improvements have been made in the past.

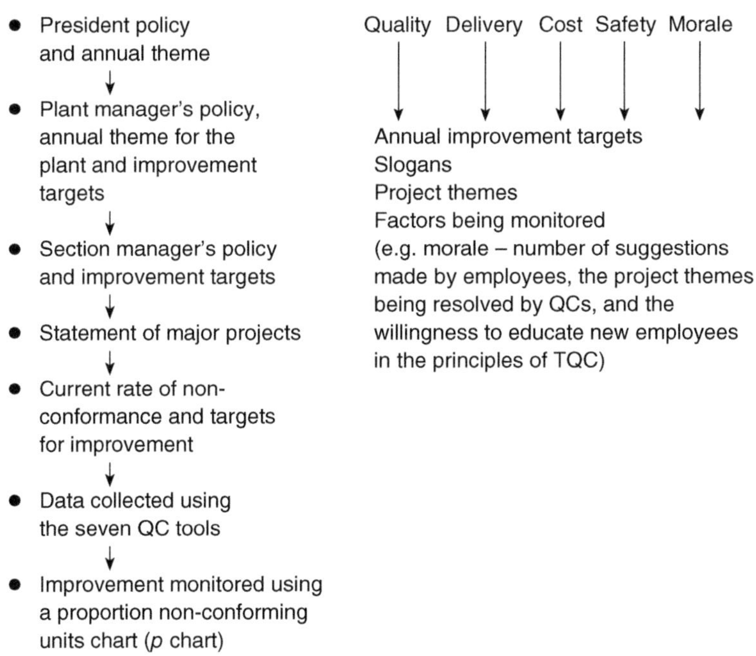

Figure 4.3 Key points of the visual display of policy deployment for a section.

Source: Dale et al. (2016).

A Check-Reflect-Improve-Scrutinise-Pass (CRISP) Approach to Policy Deployment

From an analysis of the research carried out into the policy deployment process of a 'world-class' organisation and observations of the use of policy deployment by a variety of organisations, including Japanese manufacturing companies based in the UK, the key to ensuring the effectiveness of the process is leadership and communication.

In the literature on policy deployment, there is a consistent reference to cascade and catchball as essential elements. However, based on research and practical experience, deployment is the area where management can have the most impact, but often fails to deliver what is necessary. The cascade process often does not work as expected because line managers fail to communicate effectively with their teams, perhaps due to a lack of time, skills or management style. Catchball fails to work as expected because employees do not see the results of the process, and it becomes over-reliant on individual management style. Policy deployment can be viewed in the form of a wheel (see Figure 4.4).

With business results at the hub, targets and means as the spokes, and catchball as the rim. What is required is an effective application of iterative cycles such as PDCA or similar variants (e.g. DMAIC = define, measure, analyse, improve, control or CRISP – catch, reflect, improve, scrutinise, and pass), throughout this policy deployment wheel.

CRIP stands for catch, reflect, improve and then pass the idea, which clearly embodies the catchball concept. Preferably, no one should pass on the idea immediately on receipt without first improving it. In this way, it is believed that a consensus can be achieved with maximum participation and minimal conflict. However, more importantly, it prevents

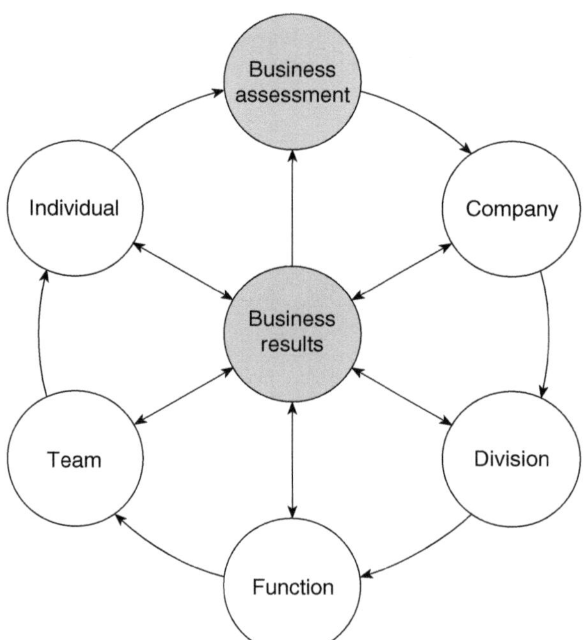

Figure 4.4 The policy deployment wheel.

'passing the buck' and, combined with PDCA, it can be used throughout the policy deployment process to monitor progress and the continued relevance of objectives. Indeed, this whole cycle can become CAPD (check, act, plan, do) when policy deployment begins with annual self-assessment and the monitoring cycle begins with the 'check' step to ensure goals remain viable and appropriate (see Akao, 1991).

One solution to the 'management' problem encountered with policy deployment is for organisations to adopt the CRISP approach. This is a development of the mnemonic CRIP, as briefly described by Mulligan et al. (1996) in relation to their discussion of HK as a strategic planning method. The addition of 'scrutinise' is an innovation which would solve the problems of unsatisfactory, incomplete policy deployment and lack of management commitment and attention to the process, which, from our research, are believed to be the main issues. Furthermore, as well as being a memorable mnemonic, the word CRISP has the added advantage of suggesting the brisk, decisive way the process is conducted.

Essentially, CRISP entails each individual and team catching the policy, reflecting and improving upon it, but, before passing the policy up and down the hierarchical chain, having their work scrutinised by the previous level to ensure the reflection and improvement are in line with the original policy (see Figure 4.5). This is the all-important 'check' aspect in the PDCA cycle, which often does not occur in an organisation's deployment process.

Although at first sight, this approach may appear to add unnecessary bureaucracy to policy deployment and extend the process timeframe, the addition of scrutiny does not have to be viewed as a burden, because the CRISP approach has the following advantages:

- Senior managers are required to demonstrate and use their leadership and communication skills.
- Senior managers can ensure that the right policy is being cascaded down the chain by checking their subordinates' planning activities.
- Senior managers can ensure that there is commonality of policy deployment throughout the organisation, that policy deployment is pervasive and that the whole organisation is pulling together as one team in the same direction (concept of alignment).
- Managers can check employees' understanding of the policy through scrutiny of their planning output and terms of reference.
- Middle and first-line managers are required to perform catchball-cascade because their teams' plans and terms of reference will be checked by senior managers and middle managers, respectively.
- Managers are required to communicate effectively in a team forum to ensure they meet senior management's scrutiny requirements.
- Managers would not be able to 'pass the policy deployment buck' without tailoring it, in conjunction with their team, because of senior management scrutiny.
- Managers would have an opportunity to demonstrate their leadership skills to senior managers throughout the process.
- Leadership and communication skill gaps can be identified, and personal development programmes organised during the process.
- Teams and individuals will see the catchball process in action when they are involved in the scrutiny of the next level's improvement of their objectives.
- Teams and individuals will 'buy in' to policy deployment when they see management using their feedback, become involved in its scrutiny, and have plans and objectives linked to company goals.
- It provides a 'closed loop' for policy deployment.

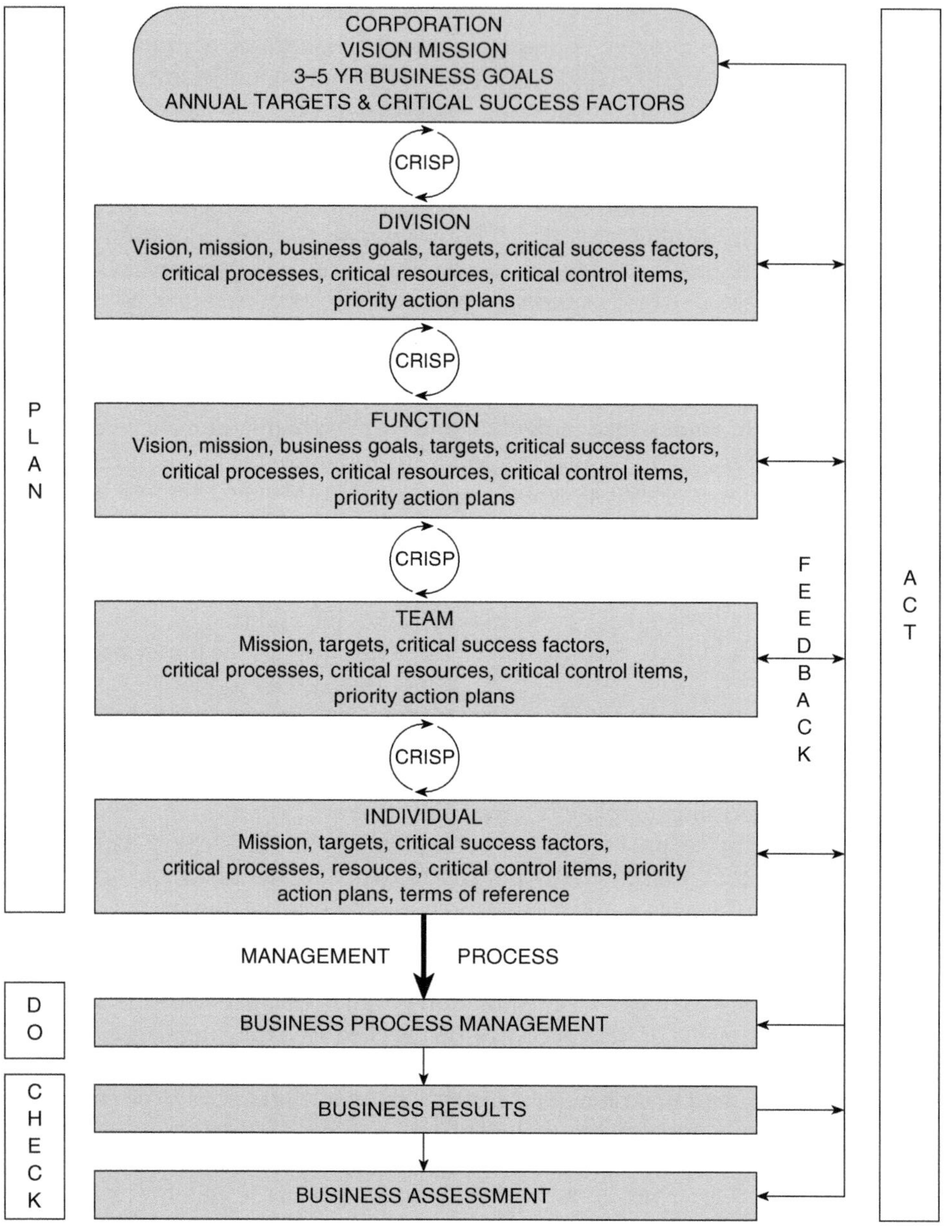

Figure 4.5 A CRISP approach to policy deployment.

The CRISP approach does not necessarily elongate the policy deployment process if it is done efficiently. Much of the work should be done pre-policy deployment, with individuals and teams working cross-functionally to draft their VFOs, plans and roles, responsibilities and objectives, after carrying out a self-assessment against a recognised excellence model, such as EFQM, Shingo Model or MBNQA. In this manner, the pre-policy deployment work will be the basis of the catchball feedback up the chain and form the framework for the final policy deployment activities. If CRISP is executed correctly, after undertaking a self-assessment, the workload during the policy deployment phase should be reduced. If the self-assessment and pre-policy deployment CRISP process has been successful, there should be no surprises or major reworks following the actual policy deployment.

An example of the application of a CRISP approach to policy deployment in a typical business unit (BU) would follow the process outlined below:

1 Before receiving the policy from the corporate organisation, following self-assessment, each team/individual in the BU would go through a cross-functional process of defining their prioritised VFO, drafting plans and their terms of reference (TOR) for the following year. This would feed the chain through the BU using CRISP. Each higher level would catch the input from the subordinate teams/individuals, reflect upon it, improve upon it by coordinating all cross-functional inputs, give the results of the improvement back to the originating teams/individuals for scrutiny to ensure that the original sense has not been subverted and then pass it on to the next level. Finally, the general manager (GM) of the BU would pass on the BU prioritised VFO to the corporate organisation for incorporation into their planning process.
2 On 'catching' the corporate organisation's VFO, the senior management team, led by the GM, would reflect upon it in relation to the BU VFO they had already determined, noting complementary and conflicting points. The senior management team would then improve the corporate organisation's VFO using the previously determined BU VFO to produce the final version. The senior management team would pass their improved and BU-specific VFO back to the corporate organisation for scrutiny to ensure that the original sense of the deployed policy had not been distorted. On confirmation from the corporate organisation that the BU VFO are acceptable, the senior management team would pass on the BU prioritised, cross-functionally co-ordinated policy – via the appropriate communication channels – to subordinate teams.
3 At the next function level, the team of line managers, led by a senior manager, would 'catch' the BU policy. The function team would reflect upon this policy, led by a senior manager, who would show where their original input to the BU policy has been used to determine the BU VFO (highlighting catchball) and how the BU VFO are linked to the corporate organisations' VFO. The team would then improve upon the BU VFO by linking their original and draft VFO to the BU VFO and determining a revised set of function-specific VFO. Again, before the line managers deployed this policy to their teams, the function-specific VFO would be passed back to the senior management team for scrutiny to ensure the interpretation and improvements were valid, there were no cross-functional conflicts and that the senior management's direction had been followed at the function level. As part of this scrutiny, the senior management team would check the function, vision, mission, goals, objectives, targets, critical processes, resources, priority action plans, critical control items, and TORs. On passing scrutiny, the function line managers would communicate the deployed policy to their teams.
4 At the team level, a similar process to that at the function level would occur, and this would be finally deployed down to an individual level as necessary.

Following the successful use of the CRISP approach to policy deployment, each individual and team should see how their original input to the planning process, following self-assessment, has been used to generate the VFO and how their role is linked to the company policy. In this way, each employee will possess TORs that they have developed with their manager and are specifically tailored to support the VFO. They will also have a vision, mission, goals, objectives, targets and means that they have contributed to and can understand in terms of their work and the company policy. Most of all, they will have played their part in policy deployment, been involved in catchball and bought into the process.

The CRISP approach is a generic management technique and can be applied to any organisation involved in a communication and feedback process. The model highlights how the PDCA cycle overlays the business process.

Policy Deployment and Two Complementary Practices

While HK is a powerful practice, the synergy with the Balanced Scorecard (BSC) and Quality Function Deployment (QFD) in achieving a coherent quality management strategy and deployment is important.

Balanced Scorecard (BSC)

The Balanced Scorecard (BSC) is a strategic management framework designed to align organisational goals with execution (Kumar et al., 2024). While HK originates from the Total Quality Management (TQM) discipline, BSC is generally presented as a performance measurement and strategic management model (Kumar et al., 2024). However, they are highly complementary and, when used together, create a robust system for achieving long-term strategic objectives while maintaining operational excellence (Dias & Tenera, 2020; Yazdi & Mennatib, 2011).

BSC provides a balanced view of performance by structuring goals and KPIs across four key perspectives as per Figure 4.6:

- Financial (e.g. revenue growth and cost management)
- Customer (e.g. customer satisfaction and loyalty)
- Internal Processes (e.g. quality improvements and operational efficiency)
- Learning & Growth (e.g. employee skills and innovation)

BSC reinforces the process of cascading goals and performance measurement by defining key performance indicators (KPIs) within each perspective. The synergy between the two lies in the way BSC provides a structured measurement framework, and HK ensures that these measures translate into concrete actions across the organisation.

HK drives strategy deployment and execution, and BSC provides a structured way to measure the organisation's success. By using both, organisations create a cohesive system where strategic quality management is both actionable and measurable. This combination ensures not only that goals are deployed but also that they deliver meaningful, measurable impact.

Quality Function Deployment (QFD)

Hoshin Kanri (HK) and Quality Function Deployment (QFD) are both structured methodologies that help organisations translate strategic objectives into actionable plans (Yazdi & Mennatib, 2011). We have explained in this chapter how HK focuses on aligning high-level strategic goals with execution synergistically. QFD ensures that customer requirements

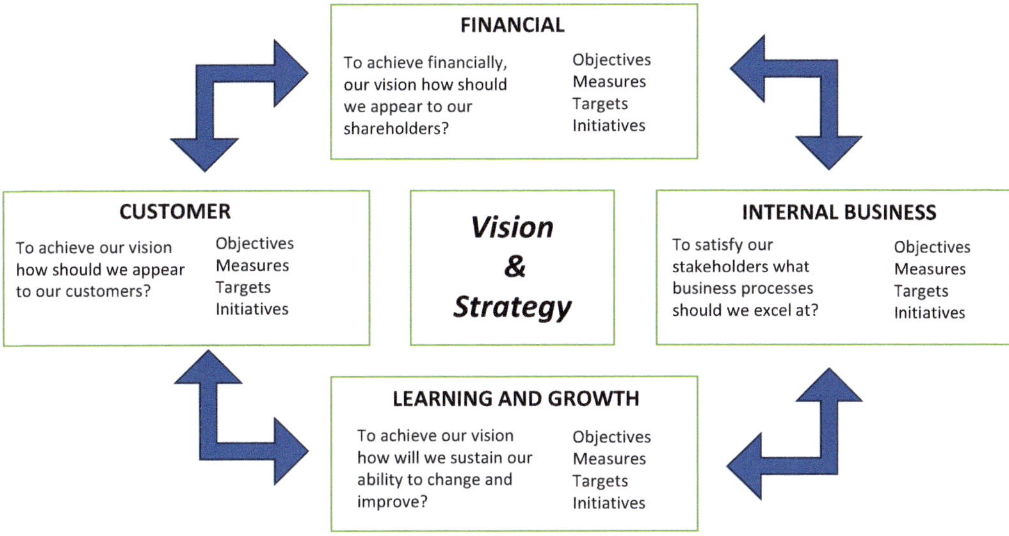

Figure 4.6 Example of the Balanced Scorecard structure.

drive the design and quality of products and services (Dehe & Bamford, 2017). QFD ensures that customer needs are translated into engineering, production, and operational requirements. It can be integrated into policy deployment to ensure that strategic objectives are directly linked to customer expectations. The X-Matrix can incorporate QFD outputs, ensuring that customer-driven priorities are embedded into strategic execution.

When used together, these frameworks create a strong customer-centric approach to strategy deployment and quality management. It creates a structured customer-driven continuous improvement loop. Combined, they form a powerful mechanism that drives customer-focused innovation, quality, and continuous improvement.

QFD will be presented in depth in a subsequent chapter (see Chapter 8).

Summary

Policy deployment is a structured planning process that uses fact-based and participative methods to deliver the achievement of objectives through continuous improvement. There are three common phases:

- Definition of the vision, mission, goals and objectives of the company.
- Deployment of the objectives throughout the organisation.
- Management of the achievement of objectives.

The following are the key concepts which underpin policy deployment:

- Clear organisational goals are understood by all employees.
- Involvement of employees in the development of action plans.
- All employees and functions are working in alignment to achieve the organisation's objectives.

- Regular review mechanism using the PDCA logic cycle, thereby promoting an environment of continuous improvement.
- Visible display of information to all concerned.

To ensure that the policy deployment process is effective, everyone in an organisation needs to have proactive involvement in the cascade and catchball process, and this should be led by the senior management team. BUs, departments and teams need to prioritise and coordinate their VFOs in an effective manner and then use these to identify and prioritise the processes which need to undergo improvements.

During the PDCA cycle of policy deployment, managers should focus on the 'check' rather than the 'act', which employees should be empowered to do. One method of ensuring that the cascade and catchball of policy deployment is effective and closed loop is to adopt the CRISP approach as described here, and encourage managers to demonstrate effective leadership and communication.

Throughout the policy deployment literature history (e.g. Dale, 1990; Akao, 1991; Jolayemi, 2008; Mothersell et al., 2008; Nicholas, 2016; Wilson et al., 2024) there is a constant emphasis on the PDCA cycle in strategic planning and daily control, the X-Matrix and the visual management aspect and the catchball and consensus principles. CRISP strengthens the catchball process, which represents the PDCA cycle between organisational levels and encourages managers to conduct policy deployment and to have their activities scrutinised before passing the improved targets and means up or down the hierarchy. At each management level, work is checked, allowing everyone to exercise and demonstrate leadership and commitment.

HK, or policy deployment, is an essential element of Quality Management and is acknowledged as an essential practice of world-class organisations (Oakland, 2011). Although policy deployment is best used by mature organisations that are a long way down the TQM or Lean Thinking road, we believe that fundamental elements such as PDCA, catchball and the X-Matrix can be used by less advanced companies to develop leadership and employee involvement and lay the foundation for future use of the full policy deployment approach.

References

Akao, Y. (1991), *Hoshin Kanri: Policy deployment for successful TQM*. Cambridge, Mass.: Productivity Press.

Butterworth, R., & Witcher, B. (2001). Realising the vision: translating strategy into action through policy management (Hoshin Kanri). *The Journal of the Institution of British Telecommunications Engineers*, 2(4), 24–30.

Dale, B., Bamford, D., & van der Wiele, T. (2016). *Managing quality: An essential guide and resource gateway*, 6th edition. UK: John Wiley & Sons.

Dale, B.G. (1990). Policy deployment. *The TQM Magazine*, December, 321–324.

Dehe, B., & Bamford, D. (2017). Quality function deployment and operational design decisions: A healthcare infrastructure development case study. *Production Planning & Control, 28*(14), 1177–1192.

Dehe, B., Bamford, D., & Kotcharin, S. (2024). Bespoke benchmarking framework employed as vehicle and platform for open innovation: A healthcare infrastructure case. *Production Planning & Control*, 35(7), 725–743.

Dias, R.M.F., & Tenera, A. (2020). Integrating balanced scorecard and Hoshin Kanri a review of approaches. *Independent Journal of Management and Production*, 11(7), 2899–2924.

Giordani da Silveira, W., Pinheiro de Lima, E., Gouvea da Costa, S., & Deschamps, F. (2017). Guidelines for Hoshin Kanri implementation: Development and discussion. *Production Planning and Control*, 28(10), 843–859.
Jolayemi, J.K. (2008). Hoshin Kanri and Hoshin process: A review and literature survey. *Total Quality Management and Business Excellence*, 19(3), 295–320.
Kondo, Y. (1998). Hoshin kanri: A participative way of quality management in Japan. *The TQM Magazine*, 10(6), 425–431.
Kumar, S., Lim, W.M., Sureka, R., Jabbour, C.J.C., & Bamel, U. (2024). Balanced scorecard: Trends, developments, and future directions. *Review of Managerial Science*, 18(8), 2397–2439.
Lee, R.G., & Dale, B.G. (1999). Policy deployment: Modelling the process. *Production Planning and Control*, 10(5), 493–501.
Marksberry, P.W. (2011). The theory behind Hoshin: A quantitative investigation of Toyota's strategic planning process. *International Journal of Business Innovation and Research*, 5(3), 347–370.
Melander, A., Löfving, M., Andersson, D., Elgh, F., & Thulin, M. (2016). Introducing the Hoshin Kanri strategic management system in manufacturing SMEs. *Management Decision*, 54(10), 2507–2523.
Mothersell, W.M., Moore, M.L., Reinherth, M.W. (2008). Hoshin Kanri planning: The system of five alignments behind the Toyota production system. *International Journal of Business Innovation and Research*, 2(4), 338–401.
Mulligan, P., Hatten, K., & Miller, J. (1996). From issue-based planning to hoshin: Different styles for different situations. *Long-Range Planning*, 29(4), 473–484.
Nicholas, J. (2016). Hoshin Kanri and critical success factors in quality management and lean production. *Total Quality Management and Business Excellence*, 27(3–4), 250–264.
Oakland, J. (2011). Leadership and policy deployment: The backbone of TQM. *Total Quality Management & Business Excellence: An Official Journal of The European Society for Organisational Excellence*, *22*(5), 517–534.
Paraschivescu, A.O., & Stoica, D.G. (2018). Catchball and total quality management. *Economy Transdisciplinarity Cognition*, 21(2), 12–17.
Santhiapillai, F.P., & Ratnayake, R.C. (2023). Lean thinking and strategy deployment: adapting Hoshin Kanri and A3-based project prioritization in police services. *The TQM Journal*, 35(8), 2503–2525.
Wilson, R., Cudney, E.A., & Marley, R.J. (2024). Current status of Hoshin Kanri. *The TQM Journal*, 36(2), 460–477.
Wood, G.R., & Munshi, K.F. (1991). Hoshin kanri: A systematic approach to break- through improvement. *Total Quality Management*, 2(3), 213–226.
Yazdi, A.K., & Mennatib, B. (2011). House of excellence: Better BSC practice through QFD plus Hoshin Kanri. *Australian Journal of Basic and Applied Sciences*, *5*(6), 1151–1159.

5 Managing Service Quality

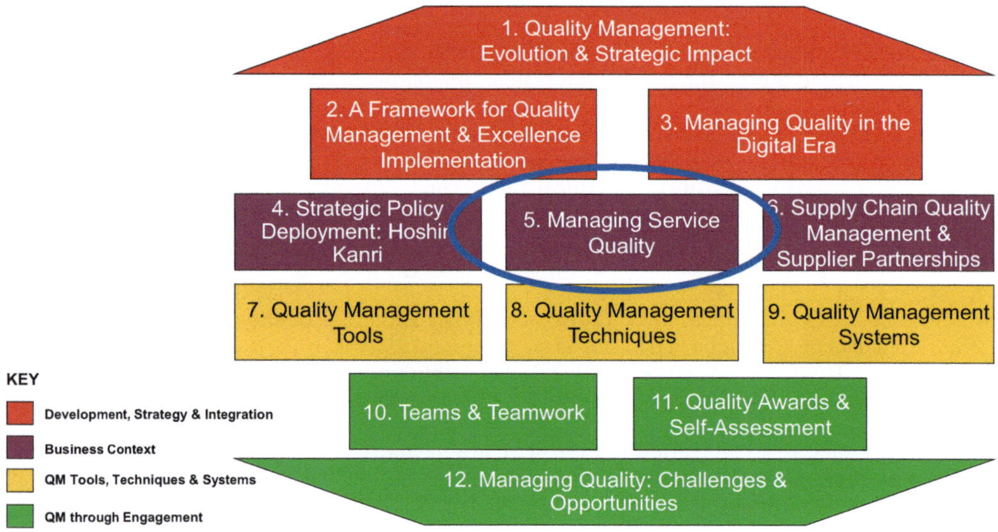

Introduction

As organisations increasingly adopt servitisation, which is defined as the transition from product-centric to service-oriented business models, the scope of quality management must extend beyond core operations, processes and tangible products (Zhang & Banerji, 2017; Rabetino et al., 2018). Unlike conventional operations, where quality is defined by measurable attributes and/or pass/fail criteria, service quality is inherently dynamic, shaped by customer interactions, expectations, and perceptions. This shift necessitates the integration of structured quality management frameworks tailored to service environments, ensuring consistency, reliability, and value creation. By embedding quality principles into both product and service offerings, organisations can enhance customer satisfaction, strengthen competitive advantage, and drive long-term business sustainability (Arora & Narula, 2018). Managing service quality is concerned with understanding what the determinants are and how they can be measured, as well as identifying the potential shortfalls in service quality and putting in place recovery strategies. Responsibility for service quality lies with operations, marketing, human resources and other management functions, working together in sync.

DOI: 10.4324/9781003594369-7

Service quality issues have been of interest for more than three decades as organisations are increasingly sourcing services that become part of their value proposition and offering bundles of services associated with their products (van Iwaarden & van der Valk, 2013; Zhang & Banerji, 2017).

Moreover, in many developed and emerging economies, the service sector now accounts for the majority of Gross Domestic Product (GDP) and employment, surpassing manufacturing and agriculture (World Trade Organization, 2022). This shift is largely due to:

- Economic maturity – as economies develop, there is a natural transition from industrial production to service-oriented industries, such as finance, healthcare, education, transport, and professional services.
- Technological advancements – digitalization, automation, and AI have enabled the growth of knowledge-based and digital services, expanding opportunities in areas like software, cloud computing, and data analytics.
- Changing consumer expectations – modern consumers demand personalised, experience-driven solutions, leading companies to integrate services with their products to enhance value (e.g., subscription models, after-sales support, and digital platforms).
- Globalisation and servitisation – many manufacturers have adopted servitisation strategies, offering maintenance, consulting, and integrated solutions to differentiate themselves in competitive markets, where the focus of business activity is increasingly on 'services' rather than 'products'.

(Bamford et al., 2015)

Services are different from products and are characterised by several aspects that have implications for the quality of service provided to customers.

- Services are typically intangible; there is usually little tangible evidence to show once a service (e.g. investment advice and consultation with a doctor) has been performed.
- The production, delivery and consumption of many services happen simultaneously. The service may not be separable from the seller/provider, and the customer may be involved in the service performance (e.g. legal advice and medical consultation). Thus, the service process, including staff at the customer interface, becomes integral to service quality.
- Because of the above points, services are heterogeneous. Variation and variability often exist in services as a function of labour inputs and non-standardisation of delivery. So, the use of quality standards in the conventional sense is more difficult.
- Finally, most services cannot be stored to meet fluctuations in demand (e.g. a doctor's appointment time, hotel rooms, and purchase of shares in a privatisation issue), so companies need to develop systems to align and manage supply and demand very effectively.

In this chapter, we first focus on the changing business environment, which has implications for service quality. The emphasis will then turn to defining service quality, measurement of service quality and the role of employees in service delivery. The final sections are concerned with the service delivery process, the need to monitor service quality, and the development of service recovery strategies.

The Service Environment

Environmental trends that impact service and quality issues relate to consumers' awareness and expectations, technological developments, and competitive elements.

Consumers, such as individuals, households or businesses, are more aware of the alternatives on offer (in relation to both services/products and provider organisations) and rising standards of service. Their expectations of service and quality are therefore elevated, and they are increasingly critical of the quality of service they experience. Expectations are what people feel a service/product should offer and what they relate to the company and its marketing mix, both the traditional elements (product, price, place, and promotion), and the extended elements of physical evidence, process and people (Singh, 2012). The physical environment includes tangible clues which might be essential (technology, infrastructure, building) or peripheral (decor, website, uniforms) to a service being bought. The service process is also critical; if systems are poor (e.g. breakdown of computer access to customer accounts in a bank, or inadequate layout), employees get blamed, and consumers perceive poor-quality service. Personnel are also integral to the production of a service and, although their degree of contact with the customer varies, all have a contribution to make, even in this enhanced technology environment.

Advancements in technology over the past decade have significantly transformed service quality, enhancing efficiency, personalisation, and accessibility. The rise of Artificial Intelligence (AI)-powered chatbots, such as those used by banks and e-commerce platforms, has improved customer service by providing instant, 24/7 support. Moreover, AI and machine learning algorithms enable companies to offer highly personalised recommendations, enhancing user satisfaction (e.g. see Amazon and Spotify). Big data analytics, widely adopted by retailers, allows businesses to anticipate customer needs and optimise supply chain operations in real time (e.g. Walmart). The proliferation of mobile apps and digital platforms, such as Uber (ride-hailing services) and Alipay (a digital payment platform and app that allows users to pay online and in person – widely used in China), has revolutionised service accessibility and convenience. Additionally, IoT-enabled predictive maintenance in industries, such as for airlines, has reduced downtime and improved reliability. These innovations have redefined service expectations, making speed, convenience, and personalisation essential dimensions of modern service quality (see Shahid Iqbal et al., 2018; Raza et al., 2020).

These advances make a significant contribution to facilitating customer–company exchanges and enhancing service levels. With computerisation, automation, robotics, and AI, organisations can greatly personalise services and increase customer loyalty. Generally, high-tech and 'high-touch' go hand in hand with better-tailored services and enhanced efficiency. Technology can free employees' time, allowing them to concentrate on the customer and improve customer–staff interaction.

In addition, the business environment is increasingly complex and competitive as a result of economic and geopolitical conditions, and increased customer choice and sophistication. Companies recognise the importance of customer service and quality, and the opportunities for attaining differentiation and achieving a competitive edge by providing superior service. Consequently, service quality is seen as a mechanism to achieve pre-eminence in the marketplace and the battle for market share, and so becomes a factor in strategic planning.

The Benefits of Good Service

Without a focus on service quality, organisations will face problems and complaints from both employees and customers and see an increase in associated costs. Furthermore, a proportion of dissatisfied customers will complain and tell others, generating adverse word-of-mouth publicity and possibly accusations of blame between personnel in the organisation, and some will switch to competitors. With a service quality programme, an organisation can expect several benefits relating to customers, employees and corporate image.

- The most frequently mentioned benefit is enhancing customer loyalty through satisfaction. Looking after present customers can generate repeat and increased business and may lead to the attraction of new customers from positive word-of-mouth communication. This is significantly more cost-effective than trying to attract new customers.
- A number of organisations highlight the additional benefit of increased opportunities for cross-selling (Yang and Ji, 2022). Comprehensive and up-to-date product knowledge and sales techniques among employees, combined with developing relationships and rapport with customers, enable staff to identify customer needs and suggest relevant services and products.
- In relation to employees, benefits may be seen in terms of increased job satisfaction, morale and commitment to the company, good employer–employee relationships, and increased staff loyalty, all of which contribute to reducing the rate of staff turnover and the associated costs of recruitment, selection and training activities. Further to this, Heskett et al. (1994) in their service–profit chain model, examined the impact of employee satisfaction and performance, employee loyalty and retention, on customer satisfaction and retention and organisational success. More recently, Sahi et al. (2024) highlighted the interdependence of internal (employee satisfaction, empowerment, and engagement) and external (customer experience and loyalty) service quality in driving success, making the service–profit chain a vital strategic tool for organisations seeking to innovate and sustain market leadership.
- In addition, good service quality enhances corporate image and may provide insulation from price competition; some customers will pay a premium for reliable service quality. Overall, successful service quality leads to reduced costs (of mistakes, operating, advertising and promotion) and increased productivity and sales, market share, profitability and business performance.

Defining Service Quality

At this point, it is useful to introduce the concept of 'service encounters' (see Grace, 2009; Pitardi et al., 2022). A service encounter refers to any interaction between a customer and a service provider that influences the customer's perception of the service. These encounters can occur through various channels, including face-to-face interactions (e.g. in a hotel or university), digital platforms (e.g. live chat support, mobile apps, or learning environment), and self-service technologies (e.g. ATMs or self-checkout kiosks). Service encounters are critical moments where service quality is experienced and evaluated, shaping customer satisfaction and loyalty. We often categorise them into three types:

- Remote encounters: interactions via digital or automated systems (e.g. using a banking app).
- Phone encounters: voice-based interactions with a service representative (e.g. call centre support).
- Face-to-face encounters: direct personal interactions (e.g. consulting a doctor).

The quality-of-service encounters are influenced by factors such as employee behaviour, responsiveness, reliability, and the emotional connection established during the interaction. Effective service encounters contribute to a positive customer experience, while poor encounters can lead to dissatisfaction and loss of trust. For example, a bank customer wishing to make an account inquiry and transaction may choose between interacting with an automated teller machine, accessing via the internet or mobile banking, or interacting with a bank employee by phone, email, or face-to-face in a branch. Every time a customer encounters any aspect of the bank and its employees, they have an opportunity to form an impression of the bank and its service. Service encounters have a high 'impact' on consumers, and the quality of the encounter is an essential element in the overall impression and evaluation of the quality of service experienced by the customer.

Service encounters also have an impact on employees in relation to their motivation, performance, job satisfaction, and rewards. Consequently, all organisations need to manage their service encounters effectively and empower their employees for the benefit of customers and staff and for the achievement of corporate goals.

Furthermore, we can witness the extent to which technology, robotics, and AI are impacting and improving service encounters (Ostrom et al., 2019; Robinson et al., 2020). More recently, attention has been given to the ability of technology to effectively customise service offerings, recover from service failure and spontaneously delight customers. It is relevant to investigate the infusion of technology as an enabler of both employees and customers in efforts to achieve these goals.

Service quality is variously defined, but essentially it has to do with meeting customer needs and requirements and with how well the service level delivered matches customers' expectations. Expectations are desires/wants, i.e., what we feel a service provider should offer, and are formed based on previous experience of a company and its marketing mix, awareness of competitors' offerings and word-of-mouth communication. Consequently, service quality becomes a consumer judgement and results from comparisons by consumers of expectations of service with their perceptions of actual service delivered (see Berry et al., 1988; Liu et al., 2015; Setiawan et al., 2020). If there is a shortfall, then a service quality gap exists which providers would wish to close. However, one needs to bear in mind that:

- Higher levels of performance lead to higher expectations.
- To find expectations greater than performance implies that perceived quality is less than satisfactory. This is not to say that service is of low quality. Quality is relative to initial expectations, one of the issues to consider when measuring service quality.

The concept of service quality gaps was first developed from the extensive research of Parasuraman, Zeithaml, Berry and their colleagues (Parasuraman et al., 1985; Zeithaml et al., 2000). They defined service quality to be a function of the gap between consumers' expectations of a service and their perceptions of actual service delivery by an organisation, and suggested that this gap is influenced by several other gaps which may occur.

Gap 1. Consumer expectations vs. management perceptions of consumer expectations

Managers' perceptions of customers' expectations may be different from actual customer needs and desires, i.e., managers do not necessarily know what customers (both internal and external) want and expect from a company. This may be remedied by market research activities (e.g. interviews, surveys, focus groups, complaint monitoring), and better communication between management and employees throughout the organisation.

Gap 2. Management perceptions of consumer expectations vs. service quality specifications actually set

Even if customer needs are known, they may not be translated into appropriate service specifications, due to a lack of resources, organisational constraints or absence of management commitment to a service culture and service quality. The need for management commitment and resources for service quality cannot be overstated.

Gap 3. Service quality specifications vs. actual service delivery

This is referred to as the service performance gap and occurs when the service that is delivered is different from management's specifications, owing to variations in the performance of personnel, employees not being able or willing to perform at a desired level. Solutions are central to human resources management and will be returned to.

Gap 4. Actual service delivery vs. external communications about the service

What is said about the service in external communications is different from the service that is delivered, i.e., advertising and promotion can influence consumers' expectations and perceptions of service. Therefore, it is important not to promise more than can be delivered (as expectations increase and perceptions decrease), and not to fail to present relevant information. Success in this area requires appropriate and timely information/communication, both internally and to external customers.

Gap 5. *Consumer expectations and perceptions of actual service*

Gaps 1 to 4 together contribute to consumers' expectations and perceptions of actual service (Gap 5). Firms need to identify the gaps prevalent in their organisation, determine the factors responsible for them, and develop appropriate solutions.

Dimensions of Service Quality

Dimensions of service quality have been researched and discussed for over 40 years. Grönroos (1984) referred to the technical and outcome quality-of-service encounters, i.e., what is received by the customer, and the functional quality of the process, i.e., the way in which the service is delivered. Typically, this includes the attitudes and behaviour, appearance and personality, service-mindedness, accessibility and approachability of customer-contact personnel. In addition, there exists the corporate image dimension of quality, which is the result of how customers perceive a company, and is built up by the technical and functional quality of its services.

One of the most widely reported sets of service quality determinants was proposed by Parasuraman et al. (1994). They suggested that the criteria used by consumers that are important in moulding their expectations and perceptions of service fit ten dimensions:

1. *Tangibles*: physical evidence
2. *Reliability*: getting it right first time, honouring promises
3. *Responsiveness*: willingness, readiness to provide service
4. *Communication*: keeping customers informed in a language they can understand
5. *Credibility*: honesty, trustworthiness
6. *Security*: physical and financial; confidentiality
7. *Competence*: possession of the required skills and knowledge of *all* employees
8. *Courtesy*: politeness, respect, friendliness
9. *Understanding*: knowing the customer, their needs and requirements
10. *Access*: ease of approach and contact

These dimensions vary with respect to how easy or difficult it is to evaluate them. Some, such as tangibles or credibility, are known in advance, but most are experience criteria and can only be evaluated during or after consumption. Some, such as competence and security, may be difficult or impossible to evaluate, even after purchase. In general, customers rely on experience when evaluating services. Subsequent factor analysis and testing by Parasuraman et al. (1988) condensed these determinants into five categories (tangibles, reliability, responsiveness, assurance and empathy) to which Grönroos (1988) added a sixth dimension, recovery.

Similarly, research about the online service quality shows that access, ease of navigation, efficiency, flexibility, reliability, personalisation, security and privacy, responsiveness, assurance and trust, site aesthetics, and price and knowledge are the main dimensions. Personalisation and personal service seem to become increasingly important, especially when problems occur or when customers must make complex decisions. AI-powered algorithms provide new solutions to this issue.

Zone of Tolerance

The zone of tolerance in service quality refers to the range of performance that customers consider acceptable, lying between their minimum acceptable level and their ideal level of service. This concept acknowledges that customers do not expect perfection in every service encounter; rather, they have a window of tolerance within which variations in service quality are seen as normal and satisfactory. Factors such as individual expectations, past experiences, and the context of the service encounter influence the boundaries of this zone.

When a service falls within this range, customers are likely to remain satisfied, even if the experience is not exemplary, as it meets the minimum threshold of acceptability.

Understanding the zone of tolerance is crucial for service providers, as it helps in managing customer expectations and prioritising improvements. By identifying the lower limit, the point at which service quality becomes unacceptable, organisations can focus their resources on eliminating these critical failures. Simultaneously, aiming for the upper limit, or the ideal level, can serve as a goal for service enhancement and differentiation. Effective management of the zone of tolerance enables companies to balance operational efficiency with customer satisfaction, ensuring that service improvements yield perceptible value without necessarily incurring unsustainable costs.

The width of the zone of tolerance varies depending on several factors, including customer expectations, service importance, and previous experiences. For instance, in a luxury hotel, guests may have a narrow zone of tolerance, expecting impeccable service with little room for variability. In contrast, for low-cost airlines, customers might tolerate minor delays or limited amenities, provided the core service – safe and affordable travel – is delivered.

Measurement of Service Quality

Measuring service quality is inherently more complex than assessing product quality due to the intangible, variable, and customer-dependent nature of services. Unlike physical goods, which can be evaluated based on standardised specifications, service quality is often assessed through customer perceptions and expectations. Effective measurement requires both objective indicators (e.g. service speed and error rates) and subjective feedback (e.g. customer satisfaction and perceived value). Common approaches include customer surveys, mystery shopping, service audits, and performance benchmarking. Organisations must also consider the dynamic nature of service expectations, meaning that quality standards must be regularly updated to reflect changing customer needs and industry benchmarks.

The measurement of service quality has also been the focus of research interest and debate over the past couple of decades. Parasuraman's dimensions of service provided the basis of the SERVQUAL questionnaire (Parasuraman et al., 1994), which was designed to measure service quality, i.e., the comparison between consumers' expectations of service (*E*) and their perceptions of actual service delivered (*P*). This is a 22-item scale with reported good reliability and validity, which can be used to better understand service expectations and perceptions of consumers. The original scale items were of the form shown in Table 5.1.

There are five key dimensions of service quality often identified and assessed:

1 Reliability – The ability to perform the promised service dependably and accurately.
2 Assurance – The knowledge, competence, and courtesy of employees, as well as their ability to inspire trust and confidence.

Table 5.1 Measuring service expectations and perceptions

Service expectations:	*e.g.*	*'Customers should be able to trust bank employees'*
(*E*)		'Banks should have up-to-date equipment'
and Perceptions:	e.g.	'I can trust the employees of my bank'
(*P*)		'My bank has up-to-date equipment'

3 Tangibles – The physical aspects of service, such as facilities, equipment, and staff appearance.
4 Empathy – The degree of personalised attention and care provided to customers.
5 Responsiveness – The willingness and ability to help customers promptly and efficiently.

During the last 30 years, a host of studies have used SERVQUAL or similar instruments to assess the dimensions of service quality in different contexts and industries (Haming et al., 2019; Pekkaya et al., 2019; Park et al., 2021).

SERVQUAL is still considered a current and relevant model for measuring service quality, even though it was developed in the 1980s, as it remains one of the most widely used instruments in the field and is frequently cited in research on service quality across various industries; however, it is important to note that some researchers may adapt or modify it to fit specific contexts and address potential limitations.

Areas of concern have focused on conceptual/theoretical, operational and interpretative issues, to include, for example:

- The disconfirmation paradigm: should service quality measurement be based on an assessment of performance minus expectations, or is attitude a better description of service quality?
- The relative focus on the process and outcome of service delivery.
- The measurement of expectations: what is being measured? Ideal, desired or adequate expectations, etc.
- The dimensionality of service quality and its applicability to all service industries and situations.
- The scaling techniques incorporated and associated importance weightings of service quality dimensions.
- Changes in attributes and importance, expectations, and perceptions over time.
- The timing of measurement: before, during, or after a particular service encounter.

Some research also incorporates the potential impact of cultural setting on the dimensionality of service quality, for example, in financial services, where Lewis et al. (2002) found eight dimensions of service quality in retail banking in Cyprus: tangibles, reliability, responsiveness, customer-contact personnel, commitment to customers, services portfolio, access, and image.

Finally, researchers such as Cottam and Lewis (2001) have considered the extent to which consumer expectations, perceptions and satisfaction may change during the course of extended service delivery and consumption. In particular, they found, from repeated measures of expectations throughout a year-long service experience, that levels of expectation, both ideal and predicted, changed during the process of service consumption, and that expectations were not a key variable in predicting satisfaction with the service.

There thus remains a considerable challenge for both academics and practitioners to refine the methods used to identify and measure appropriate dimensions of service quality and the gaps that exist for organisations.

The Kano Model

The Kano model is a framework that helps organisations understand how different service attributes influence customer satisfaction (Madzík, 2018). Unlike other quality models that may assume a linear relationship between service performance and satisfaction, the Kano model highlights that customer expectations vary, and not all service improvements lead to increased satisfaction in the same way (Materla et al., 2019). This perspective is particularly useful in the context of service quality, where customer experiences are shaped by both tangible and intangible factors, such as responsiveness, personalisation, and engagement.

The model categorises service attributes into different types: basic, must-have or must-be factors, performance or one-dimensional factors, and attractive, excitement or delight factors. Basic or must-be factors are fundamental expectations that customers take for granted – such as cleanliness in a hotel or secure payment processing in an online service. Failing to meet these leads to dissatisfaction, but exceeding them does not significantly enhance satisfaction. Performance or one-dimensional factors have a direct, proportional impact on customer satisfaction, such as reduced waiting times in a call centre or faster food delivery. The better these attributes are, the more satisfied customers become. Attractive or excitement factors, however, are unexpected features that create delight, such as a free room upgrade at a hotel or proactive problem resolution by customer support. However, the absence of attractive factors would not negatively impact customer satisfaction. Hogstrom et al. (2010) and Materla et al. (2019) also mentioned the reverse and indifferent factors; they explained that the absence of reverse attributes leads to customer satisfaction and that indifferent attributes do not have any impact on customer satisfaction or dissatisfaction, as illustrated in Figure 5.1.

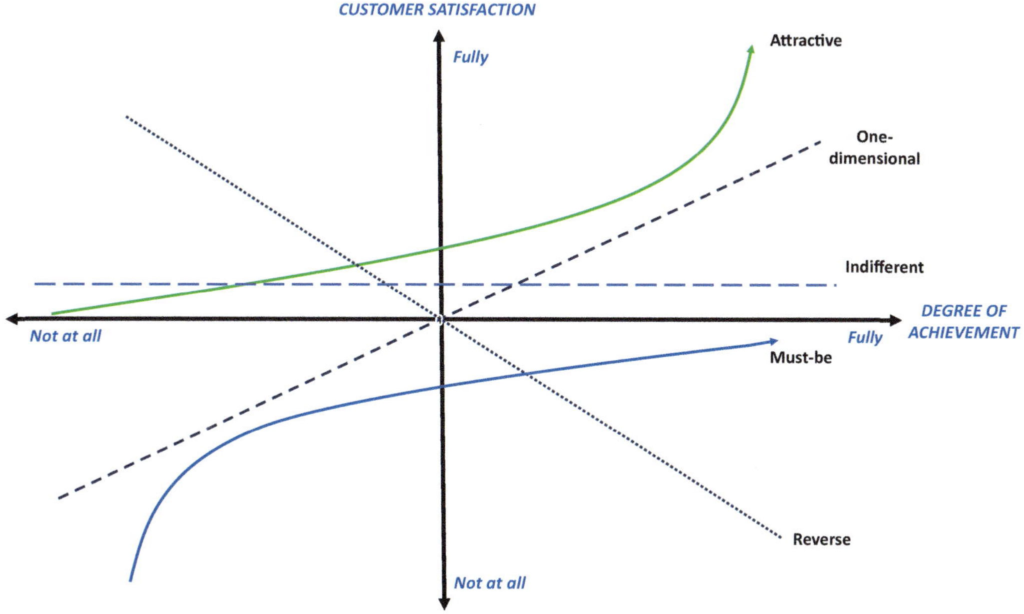

Figure 5.1 Kano model.

(Adapted from Materla et al. (2019)).

In service industries, understanding and balancing these factors is key to designing experiences that not only meet expectations but also create memorable and differentiated services.

Human-Centricity in Service Delivery

Having assessed customer needs, organisations must set standards/specifications and systems for service delivery to include the relevant dimensions of customer service, i.e., to avoid Gap 2. This implies a requirement for management commitment to a service culture and service quality, and for the allocation of appropriate resources – in relation to products, systems, environment, and people.

The subsequent challenge is to ensure that the service delivered meets the specifications set. This depends on the performance of *all* employees, who must be able and willing to deliver the desired levels of service. Employees' contributions in meeting customer needs and, thus, influencing customer perceptions of service cannot be overstated. Success depends on the development of a human-centric system and enlightened employee policies for recruitment and selection, training, motivation and rewards for all employees – both customer-contact and back-office staff.

Internal Marketing

Internal marketing (IM) views employees as internal customers and jobs as internal products, and a company needs to sell its jobs to employees before selling its service to customers, satisfying the wants of internal customers and upgrading the capability to satisfy the needs of external customers (Ahmed & Rafiq, 2003). The core idea is that satisfied and committed employees lead to better customer experiences, ultimately improving organisational performance. IM involves communication, training, empowerment, and reward systems to align employees with the company's goals and enhance their service mindset.

The concept of IM has been researched by Rafiq & Ahmed (2000), who studied the origins, nature, scope and application of this human-centric idea and considered how it might be developed to take greater account of the social and non-economic needs and interests of people working in an organised enterprise. Furthermore, Ahmed & Rafiq (2002) discussed the development and evolution of the IM concept, models of IM, and links with service quality, customer satisfaction, customer loyalty, and profitability. Over time, IM evolved into a broader strategy, influencing areas like employee engagement, organisational culture, and service quality improvement. Today, it is applied in various industries, particularly those where customer experience is a key differentiator.

People and Culture Policies

People and culture policies refer to the formal guidelines and frameworks that govern employee management within an organisation. These policies cover key areas such as recruitment, training, compensation, performance management, employee engagement, and workplace culture. Personnel and employee policies contribute to the quality of service delivered to the final customer. This includes relationships between: customer-contact and back-office employees; operations and non-operations staff; and staff and management at all levels and locations.

Successful people and culture policies include the recruitment and selection of the 'right' people. Key characteristics for employees to perform effectively may relate to: process and technical skills; interpersonal and communication skills; flexibility and adaptability; and empathy with the customer. It is also vital to identify the training needs of new and present employees with respect to technical and interpersonal dimensions, and to consider employment conditions, i.e., employees' wants and attitudes regarding working conditions, benefits and welfare. This is undertaken, typically, via a training audit. Subsequently, training programmes may be developed to provide product, company and systems knowledge and also interpersonal and communication skills. In relation to Gap 3, experts indicate that success will depend on:

- *Teamwork*: evidenced by caring management and involved and committed employees.
- *Employee-job fit*: the ability of employees to perform a job.
- *Technology–job fit*: are the 'tools and technologies' appropriate for the employee and the job?
- *Perceived control*: do employees have flexibility in dealing with customers? If not, stress levels may rise and performance decrease.
- *Supervisory control systems*: based on behaviours rather than 'output quality'.
- *Avoidance of role conflict*: for employees in satisfying their expectations of the company *and* the expectations of customers.
- *Avoidance of role ambiguity*: employees should know what is expected of them and how performance will be evaluated and rewarded.

Customer service training programmes are typically designed to move a company to a service-oriented culture by breaking down barriers and improving internal communications. Advantages are seen to be: creating an atmosphere of all working towards a common goal; understanding the work of others; and encouraging all staff to have responsibility and authority for achieving corporate objectives, which includes empowering employees to exercise judgement and creativity in responding to customers' needs.

Employees also need to be supervised, and systems set to monitor and evaluate their performance (e.g. product knowledge tests and mystery shoppers) and satisfaction. In addition, organisations have a variety of recognition and reward schemes for excellent employees, awards may be financial or not and may involve career development.

Service Delivery

In relation to service delivery, organisations need to avoid Gap 4, i.e., a failure to deliver the service as promised. Once a company has successfully assessed customer needs, translated them into service systems and standards, and recruited and trained employees, it must then manage its 'promises'. Quality systems are of benefit in helping to ensure this. Further, a company needs appropriate advertising and promotion so that the service that is offered in external communications matches the service that the organisation can deliver. Advertising and promotion affect customers' expectations and perceptions of the delivered service, so it is important not to promise more than can be delivered. Realistic communications are needed so as not to increase expectations unnecessarily and decrease perceptions of quality.

Monitoring Service Quality

A critical element in any service strategy is for a company to have in place systems to measure and monitor success. These include research and evaluation among employees and customers, using focus groups, discussions, surveys and interviews, and sometimes mystery shoppers and 'control' branches. Collection and analysis of customer complaints is very valuable, and for some organisations, key indicators are provided by service guarantees and recovery activities.

Many, if not most, service providers make promises and offer guarantees with respect to products/services, delivery and aspects of performance. In the private sector, these may be an element in a company's competitive armoury, for example: hotels which offer cash compensation or free accommodation if issues are not resolved in 30 minutes; a pizza delivery which becomes free after a certain time delay; and a retailer which publishes promises to customers, including information on consumer rights with respect to faulty goods and the return of goods. For example, Utility companies have been actively developing and promoting, for competitive and customer-centric reasons, their service guarantees. United Utilities publishes standards for all its services and makes various promises. For instance, with respect to planned mains repairs, they make promises relating to the provision of information, timing and restoration of water supply, and cash compensation is available, and levels clearly specify if these promises are not met.

Service Failure

To turn to service failures and customer complaints, service providers now appreciate that only a small proportion of dissatisfied customers complain, so that complaint data are not a true reflection of the extent of customer dissatisfaction (Knox & Van Oest, 2014). The reasons why dissatisfied customers choose not to complain were originally discussed by Horovitz (1990) and include:

- Fear of hassle or too much trouble to complain.
- No one is available to complain to, or there is no easy channel by which to communicate disquiet.
- No one cares, and it won't do any good.
- Do not know where to complain.
- Customers see themselves as a source of service problems due to their failure to perform in the creation of the service.

Organisations should strive for zero defects in their service delivery – to get things right the first time. Consequently, many companies develop service quality systems that tend to be rigid, with sophisticated techniques and structured personnel policies, to try to provide consistent high-quality service. However, all service organisations will find themselves in situations where failures occur in their encounters with customers with respect to one or more dimensions of service quality, and where they need to deal with customer dissatisfaction and complaints (Yilmaz et al., 2016). Problems do occur (e.g. bad weather may delay an airline flight, or employees may be sick and absent), and mistakes will happen (e.g. a hotel room not ready on time, a dirty rental car, and a lost suitcase).

One should note, however, that a service failure may not only relate to a flawed outcome. In fact, Tan et al. (2021) define a service failure as 'a service performance that falls below the expectation of one or a few customer(s)'. Several researchers have investigated service failures and many have attempted to classify them (e.g. Bamford & Xystouri, 2006) in relation to problems in the service organisation (e.g. about employees, equipment, technology, and systems); those which may be customer-induced, and those that are a result of the actions of other organisations. In addition, there is evidence that service failure can lead to a decline in employee morale and service performance (Bamford & Xystouri, 2006).

Service Recovery

The actions that a service provider takes to respond to service failures are referred to as service recovery (Grégoire & Mattila, 2021). Service recovery is defined by Knox and Van Oest (2014) as 'the set of actions a company takes in response to the failure to keep the customer's goodwill'. The response from an organisation to service failures needs to be the result of a conscious, coordinated effort of the firm to anticipate that service flaws will occur, and to develop procedures, policies, and human competencies to deal with them.

When something does go wrong, what do customers expect from the service firm? Zemke (1994) concluded that customer expectations for service recovery are: to receive an apology for the fact that the customer is inconvenienced; to be offered a fair solution for the problem; to be treated in a way that suggests the company cares about the problem, about fixing the problem, and about the customer's inconvenience; and to be offered value-added compensation for the inconvenience.

Critical to the service recovery process is the empowerment of front-line employees. It is essential to give personnel the authority, responsibility and incentives to identify, care about and solve customer problems and complaints; to allow them to use their initiative and judgement to respond flexibly, and to act with respect to the best solutions to satisfy customers. Empowerment is seen to lead to better job performance and improved morale. It is a form of job enrichment, evidenced by increased commitment to jobs and reflected in attitudes towards customers. Knowing that management has confidence in employees helps to create positive attitudes in the workplace and good relationships between employees and between employees and customers.

Lewis and Spyrakopoulos (2001) investigated service failures and service recovery strategies used by banks to respond to them. A survey was developed to measure customers' perceptions of the magnitude of service failures and the effectiveness of service recovery strategies. Service failures were related to banking procedures, mistakes, employee behaviour and training, technical failures and omissions of the banks. They were found to vary in importance, and some were more difficult to deal with satisfactorily than others. Different service recovery strategies (e.g. corrections, compensation, apologies, and explanations) were more effective for failures. Further, customers with long relationships and higher deposits with their banks were more demanding with respect to service recovery.

An interesting stream of research has emerged in response to the question 'how do customers evaluate companies' service recovery strategies?'. Perceived justice comprises three elements (see Yilmaz et al., 2016):

- *Distributive justice*: the perceived fairness of the outcome. What did the offending firm offer the customer to recover from the service failure (e.g., refund or replacement)?
- *Interactive justice*: this refers to the perceived fairness of the way the customer is treated during the complaint-handling process and includes courtesy and politeness of

personnel, empathy, effort in resolving the situation and the firm's willingness to apologise and provide an explanation for the service failure.
- *Procedural justice*: the perceived fairness of the process used to rectify a service failure, e.g., the speed (or delay) in processing and correcting complaints, the accessibility and flexibility of the procedures.

The main conclusion to be drawn from the available evidence is that, to recover effectively from service failure, an organisation must provide a fair outcome, with a sincere apology, while taking the blame and acting swiftly to recover from the failure. Further, interactions between the justice dimensions mean that failure to deliver on one of them can negatively impact the total success of the recovery. The effectiveness of service recovery strategies has also been shown to depend on factors such as the service type, the magnitude of the failure, the type of failure (i.e. process or outcome), prior experience, service recovery expectations, purpose of purchase and attitudes towards complaining.

Overall, from a review of the available research evidence, it is possible to suggest several requirements for companies to create an effective service recovery programme:

- Seek out possible failure points. Focus on critical service encounters and try to anticipate problems, in particular as few people complain.
- Educate customers and encourage complaints. Make it easy to complain and offer service guarantees.
- Monitor service process, detect, and track service failures, and analyse complaint data. Measure performance against company standards. Use the findings to improve service quality and prevent failures and dissatisfaction from happening again.
- Engage in customer research, focused on both the process and outcome of service delivery, to identify when things go wrong; track how service recovery is implemented; and measure again to see if the recovery has satisfied the customer (i.e. follow-up feedback).
- Develop proactive service recovery, to include, in addition to reinstatement, elements such as: initiation of the recovery process, which can enhance customers' evaluation of the service provider; apologies, to show that the customer's problem is being taken seriously and is important to the organisation; speedy response; and recompense as appropriate.
- Train, empower and facilitate employees to recover. Service recovery typically involves interpersonal interaction and communication skills, and knowledge skills, with implications for human resources management.
- Show management commitment. Employees need to be supported by senior management that commits to absolute customer satisfaction.

Summary

The business environment is characterised by changing customer expectations, technological and product advances, legislative and political developments, and economic and competitive conditions, which contribute to an increasing emphasis on service quality for all organisations, in both the services and manufacturing sectors. Managing service quality necessitates an integrated approach from operations, marketing, human resources and other key managers/areas of a business.

Organisations need clearly defined service strategies with top management commitment and leadership. They need to understand their service encounters (both internal and external to the company) and potential failure points, and to avoid service quality shortfalls or gaps. This can be achieved by researching both service personnel and customers, identifying key dimensions of service quality, and developing appropriate service quality initiatives. Successful service strategies will include emphasis on products/services, delivery systems and procedures, technology, and personnel – their skills and commitment to the organisation and its customers.

The outcome of a successful service strategy will be satisfied and retained employees and customers, with consequent benefit to the organisation. The links between customer retention and profitability have been evident for several years. There is a growing awareness of the 'lifetime value' of retained customers in terms of the revenues and contributions earned from a long-term relationship, i.e., the longer the association between company and customer, the more profitable the relationship for the company.

References

Ahmed, P.K., & Rafiq, M. (2002). *Internal marketing: Tools and concepts for customer focused management*. Oxford: Butterworth Heinemann.

Ahmed, P.K., & Rafiq, M. (2003). Internal marketing issues and challenges. *European Journal of Marketing*, 37(9), 1177–1186.

Arora, P., & Narula, S. (2018). Linkages between service quality, customer satisfaction and customer loyalty: A literature review. *IUP Journal of Marketing Management*, 17(4), 30.

Bamford, D., Moxham, C., Kauppi, K., & Dehe, B. (2015). *Going the distance: Sport operations management in the public and third sectors*. Public Sector Operations Management, Routledge, ISBN 978-1-13-881369-4.

Bamford, D., & Xystouri, T. (2006). Recovering from service failure: Generating profit through greater customer satisfaction. *Strategic Direction*, 22(6), 37–39.

Berry, L.L., Parasuraman, A., & Zeithaml, V.A. (1988). The service-quality puzzle. *Business Horizons*, July/August, 35–43.

Cottam, A.M., & Lewis, B.R. (2001). *The measurement of expectations: Timing and relevance issues in services consumption*. Manchester: Manchester Business School.

Grace, D. (2009). An examination of consumer embarrassment and repatronage intentions in the context of emotional service encounters. *Journal of Retailing and Consumer Services*, 16(1), 1–9.

Grégoire, Y., & Mattila, A.S. (2021). Service failure and recovery at the crossroads: Recommendations to revitalize the field and its influence. *Journal of Service Research*, 24(3), 323–328.

Grönroos, C. (1984). *Strategic management and marketing in the service sector*. UK: Chartwell-Bratt.

Grönroos, C. (1988). Service quality: The six criteria of good perceived service quality. *Review of Business*, 9(3), 10–13.

Haming, M., Murdifin, I., Syaiful, A.Z., & Putra, A.H.P.K. (2019). The application of SERVQUAL distribution in measuring customer satisfaction of retails company. *Journal of Distribution Science*, 17(2), 25–34.

Heskett, J.L., Jones, T.O., Loveman, G.W., Sasser, W.E., & Schlesinger, L.A. (1994). Putting the service profit chain to work. *Harvard Business Review*, 72(2), 164–174.

Hogstrom, C., Rosner, M., & Gustafsson, A. (2010). How to create attractive and unique customer experiences: An application of Kano's theory of attractive quality to recreational tourism. *Marketing Intelligence & Planning*, 28(4), 385–402.

Horovitz, J. (1990). *Winning ways: Achieving zero defect service*. Cambridge, Mass.: Productivity Press.

Knox, G., & Van Oest, R. (2014). Customer complaints and recovery effectiveness: A customer base approach. *Journal of Marketing*, 78(5), 42–57.

Lewis, B.R., Ioannou, M., & Cui, C.C. (2002). *Service quality in the cypriot banking sector: Determinants and gaps*. Manchester: Manchester Business School.

Lewis, B.R., & Spyrakopoulos, S. (2001). Service failures and recovery in retail banking: The customers' perspective. *International Journal of Bank Marketing*, 19(1), 37–47.

Liu, W., Xie, D., Liu, Y., & Liu, X. (2015). Service capability procurement decision in logistics service supply chain: A research under demand updating and quality guarantee. *International Journal of Production Research*, 53(2), 488–510.

Madzík, P. (2018). Increasing accuracy of the Kano model: A case study. *Total Quality Management & Business Excellence*, 29(3–4), 387–409.

Materla, T., Cudney, E.A., & Antony, J. (2019). The application of Kano model in the healthcare industry: A systematic literature review. *Total Quality Management & Business Excellence*, 30(5–6), 660–681.

Ostrom, A.L., Fotheringham, D., & Bitner, M.J. (2019). Customer acceptance of AI in service encounters: Understanding antecedents and consequences. *Handbook of Service Science*, II, 77–103.

Parasuraman, A., Zeithaml, V.A., & Berry, L.L. (1985). A conceptual model of service quality and its implications for future research. *Journal of Marketing*, 49 (Fall), 41–50.

Parasuraman, A., Zeithaml, V.A., & Berry, L.L. (1988). SERVQUAL: A multiple item scale for measuring consumer perceptions of service quality. *Journal of Retailing*, 64(1), 14–40.

Parasuraman, A., Zeithaml, V.A., & Berry, L.L. (1994). Reassessment of expectations as a comparison standard in measuring service quality: Implications for further research. *Journal of Marketing*, 58(January), 111–124.

Park, S.J., Yi, Y., & Lee, Y.R. (2021). Heterogeneous dimensions of SERVQUAL. *Total Quality Management & Business Excellence*, 32(1–2), 92–118.

Pekkaya, M., Pulat İmamoğlu, Ö., & Koca, H. (2019). Evaluation of healthcare service quality via Servqual scale: An application on a hospital. *International Journal of Healthcare Management*, 12(4), 340–347.

Pitardi, V., Wirtz, J., Paluch, S., & Kunz, W.H. (2022). Service robots, agency and embarrassing service encounters. *Journal of Service Management*, 33(2), 389–414.

Rabetino, R., Harmsen, W., Kohtamäki, M., & Sihvonen, J. (2018). Structuring servitization-related research. *International Journal of Operations & Production Management*, 38(2), 350–371.

Rafiq, M., & Ahmed, P.K. (2000). Advances in the internal marketing concept: Definition, synthesis and extension. *Journal of Services Marketing*, 14(6), 449–462.

Raza, S.A., Umer, A., Qureshi, M.A., & Dahri, A.S. (2020). Internet banking service quality, e-customer satisfaction and loyalty: The modified e-SERVQUAL model. *The TQM Journal*, 32(6), 1443–1466.

Robinson, S., Orsingher, C., Alkire, L., De Keyser, A., Giebelhausen, M., Papamichail, K.N., … & Temerak, M.S. (2020). Frontline encounters of the AI kind: An evolved service encounter framework. *Journal of Business Research*, 116, 366–376.

Sahi, G.K., Mahajan, R., Jaiswal, A.K., & Patel, P.C. (2024). The strongest link: Service profit chain as a conduit for enabling entrepreneurial orientation and multidimensional service performance. *IEEE Transactions on Engineering Management*, 24(6), 40–45.

Setiawan, E.B., Wati, S., Wardana, A., & Ikhsan, R.B. (2020). Building trust through customer satisfaction in the airline industry in Indonesia: Service quality and price fairness contribution. *Management Science Letters*, 10(5), 1095–1102.

Shahid Iqbal, M., Ul Hassan, M., & Habibah, U. (2018). Impact of self-service technology (SST) service quality on customer loyalty and behavioral intention: The mediating role of customer satisfaction. *Cogent Business & Management*, 5(1), 1.

Singh, M. (2012). Marketing mix of 4P's for competitive advantage. *IOSR Journal of Business and Management*, 3(6), 40–45.

Tan, T.M., Balaji, M.S., Oikarinen, E.L., Alatalo, S., & Salo, J. (2021). Recover from a service failure: The differential effects of brand betrayal and brand disappointment on an exclusive brand offering. *Journal of Business Research*, 123, 126–139.

van Iwaarden, J., & van der Valk, W. (2013). Controlling outsourced service delivery: Managing service quality in business service triads. *Total Quality Management and Business Excellence*, 24, 9–10.

World Trade Organization. (2022). The future of trade lies in services: key trends. Available at https://www.wto.org/english/res_e/booksp_e/trade_in_serv_devpt_chp1_e.pdf

Yang, G., & Ji, G. (2022). The impact of cross-selling on managing consumer returns in omnichannel operations. *Omega*, 111, 102665.

Yilmaz, C., Varnali, K., & Kasnakoglu, B.T. (2016). How do firms benefit from customer complaints?. *Journal of Business Research*, 69(2), 944–955.

Zeithaml, V.A., Parasuraman, A., & Malhotra, A. (2000), *A conceptual framework for understanding e-service quality: Implications for future research and managerial practice*. Report 00–115. Cambridge, Mass: Marketing Science Institute.

Zemke, R. (1994). Service recovery. *Executive Excellence*, 11(9), 17–18.

Zhang, W., & Banerji, S. (2017). Challenges of servitization: A systematic literature review. *Industrial Marketing Management*, 65, 217–227.

6 Supply Chain Quality Management and Supplier Partnerships

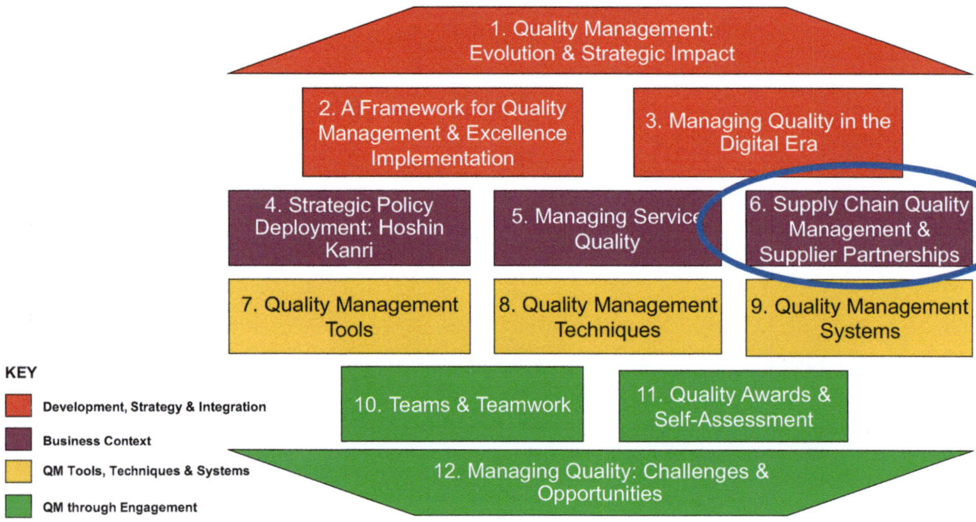

Introduction

By considering the supply chain in this chapter, we are reflecting on how an organisation can manage quality beyond its boundaries. Procured resources can be physical, such as parts, components, or products, but they can also be labour, services, and information. In fact, Heinis et al. (2022) reported that the importance of the procurement of services, which has increased, now represents a significant proportion of most organisations' external expenses, in many cases surpassing the share spent on materials. The quality of purchased resources is therefore crucial to an organisation and consequently to its success in the marketplace. As outsourcing has become the norm, bought-in components and services can be responsible for some 70 to 80 per cent of the final cost and overall quality of a company's commercial offering. It is therefore clear that suppliers and contractors are critical to the quality, competitiveness and the overall performance of the focal firm. Many major companies have, during the last 25 years or so, started to encourage their suppliers to develop and align their quality management systems and programmes. For example, it is very common for an organisation to promote and encourage its suppliers to adopt lean, six sigma, and other continuous improvement philosophies, to eliminate non-value-added activity,

DOI: 10.4324/9781003594369-8

improve manufacturing operations, and become more flexible and responsive (Cudney & Elrod, 2010; So & Sun, 2010; Minshull et al., 2022). This is referred to as the concepts of Supply Chain Quality Management (SCQM) and Supply Chain Quality Integration (SCQI) (Fernandes et al., 2017; Luo et al., 2023). SCQM enables companies to expand their quality management beyond their boundaries (Alkalha et al., 2019). Alkalha et al. (2024) also explain that SCQI is an advancement of SCQM and is concerned with coordinating and aligning intra- and inter-organisational quality systems to produce and deliver high-quality products and services. To become and stay competitive, organisations have to (i) develop long-term partnerships with their suppliers and (ii) develop their suppliers' capabilities.

The process of focal firms working together with their suppliers to enhance their processes' quality, innovation, and performance is enabled by coherent supplier development programmes, requires effective and efficient supplier relationship management strategies, and long-term partnership development. In this chapter, we will discuss industry-wide initiatives and best practices to facilitate closer relationships between buyers (from the focal firm) and suppliers to improve quality across the value chain and, as a byproduct, innovation, sustainability, and resilience of the supply network (Siagian et al., 2021). Specifically, this chapter examines the key issues in developing long-term partnerships.

Supply Chain Quality Management and Integration

SCQM refers to the systematic coordination of quality-related activities across the supply chain to ensure that products and services meet or exceed customer expectations (Zimon et al., 2022). It extends traditional quality management principles, such as continuous improvement, customer focus, and process orientation, beyond organisational boundaries to include suppliers, logistics providers, and other partners. Its emphasis is on managing quality at every link in the chain, with a strong focus on preventing defects rather than reacting to them (Alkalha et al., 2019).

SCQI, on the other hand, represents a more advanced and strategic level of quality collaboration within the supply chain (Yu & Huo, 2018; Alkalha et al., 2024, 2025). While SCQM may involve managing quality across partners, SCQI focuses on embedding quality practices, processes, and systems seamlessly between the buyer and supplier organisations. This includes aligning quality objectives, sharing data in real time, co-developing products or processes, and using integrated performance measurement systems. SCQI reflects a high degree of interdependence and trust, where quality is not just managed across entities but is integrated into the fabric of cross-organisational operations.

Effective buyer–supplier development and partnerships are fundamental enablers of both SCQM and SCQI (Kim et al., 2016). Partnerships facilitate coordinated quality activities, such as joint quality planning, problem-solving, and performance monitoring. A collaborative relationship allows buyers to work constructively with suppliers to ensure compliance with standards, resolve quality issues proactively, and promote continuous improvement throughout the supply chain. Trust and open communication are essential here, as they enable the timely exchange of information and a shared understanding of quality expectations (Yu & Huo, 2018). In the case of SCQI, the role of partnership becomes even more critical and strategic. Here, the relationship moves beyond coordination to full integration, where both parties align their quality objectives, share systems and data, and engage in joint innovation. Such integration requires a high level of mutual trust, transparency, and commitment, as well as shared investments in technology, processes, and people (Zhang & Huo, 2013). A strong partnership reduces transactional behaviours and fosters a relational governance model where both firms are jointly accountable for quality outcomes. Without such a partnership, attempts at SCQM and SCQI are likely to be superficial, fragmented, or unsustainable.

Long-Term Buyer–Supplier Partnership

The traditional, open-market bargaining approach to buyer–supplier dealings has been based on the assumption that the parties involved are adversaries who have conflicting objectives and are engaged in a win-lose contest, based upon tough negotiations, price orientation and cost undercutting with a short-term perspective and time horizon (Cousins et al., 2008). This approach, which in reality very few firms can really benefit from, focuses on negative issues, involves power abuse, and is characterised by uncertainty, all of which can seriously undermine, rather than reinforce, the competitiveness of both buyers and suppliers.

Partnership demands a deeper form of relationship. It means working together towards common aims and aspirations. It is based on the principle that both parties can gain more through co-operation than through conflict. Partnerships are characterised by mutual trust and commitment, integrity, integration, co-operation, honesty, a willingness to openly declare issues and work together to solve problems, the sharing of data and ideas, improvements and best practices (Kotcharin et al., 2024). To achieve this level of collaboration, firms need to clearly understand their responsibilities, focus on open innovation in Research & Development (R&D), and share a desire to continuously improve the quality of their products and services. As an example, most first-tier suppliers are no longer being handed a prescriptive design by the original equipment manufacturer (OEM). Instead, they are provided with a specification which the component, when assembled, must conform to. Other examples include suppliers setting up a manufacturing or distribution facility alongside their customers and a supplier's staff working in the customer's plant or production site (e.g. DHL and Jaguar Land Rover (JLR)). In many respects, this form of partnership has similarities with a vertically integrated supply chain model, but without the difficulties of managing a complex business across different types of technologies and processes. One of the drivers for closer relationships with suppliers has been the move by OEMs to concentrate on their core competencies and to shift other activities and responsibilities to their suppliers while enhancing performance. It must not be forgotten that a partnership does not come about by accident, and it cannot be sustained by inattentiveness. It is a management process which needs to be carefully designed, orchestrated and nurtured (Fawcett et al., 2012; Touboulic & Walker, 2015).

To develop a viable long-term business relationship, considerable changes in behaviour and attitude are required and need to be promoted in both buyer and supplier organisations. Focal firms need to be prepared to develop plans and procedures for working with their suppliers and commit resources to this. For their part, supplier firms must accept full responsibility for the quality of their shipped product and not rely on the buyer's receiving inspection to assess if it meets their requirements. As a prerequisite of partnership, both parties have to reach an agreement on how they will work together, what they want to achieve from the relationship and how to resolve any problems which may arise. To ensure that the partnership is sustainable, it is important that the objectives of the agreement are examined and discussed on a regular basis.

The typical benefits of developing a long-term business partnership include:

- Reduction and elimination of the inspection of supplied parts and materials.
- Improved product and service quality, and delivery performance and responsiveness.
- Improved productivity, increased stock turns and lower inventory carrying costs and reduced costs per piece.
- Enhanced security and stability of supplies.
- Promotion of the transfer of ideas, expertise and technology between the firms and dissemination and implementation of best practice.
- Lead to joint problem-solving activities, with the focal firm assisting the supplier to help improve processes.
- Integration of business practices and procedures between buyer and supplier.
- A comprehensive buyer–supplier communications framework to ensure the supplier is provided with early access to the future designs and business plans and is kept informed of changing customer requirements.
- Supplier's contribution to the buyer's design process, undertaking development work and monitoring technological trends, leading to innovative products and services, and other business opportunities.
- Development of sustainable growth in terms of investment in equipment and manufacturing/business resources.
- Exposure of the supplier to new tools, techniques, systems and business practices.
- Provision by the buyer of an advisory service to suppliers in terms of training, equipment and operating methods.

Barriers to Developing Buyer–Supplier Partnerships

Developing partnerships is not without difficulties. Dale et al. (2016) reported that certain aspects of the buyer–supplier relationship can act as a barrier to supplier development. These include:

- Poor communication and feedback, and a lack of data sharing
- Supplier complacency
- Misalignment or misguided supplier improvement objectives
- Lack of buyer credibility as viewed by their suppliers
- Misconceptions regarding purchasing power

Poor Communication and Feedback, and a Lack of Data Sharing

Communication, feedback and data sharing between a focal organisation and its suppliers is not always optimum. Sometimes it can even be inadequate and ineffective, and parties would not directly realise the negative consequences. The main dissatisfactions expressed by suppliers relate to technical specifications and requirements, the lack of consultation on design and product engineering issues and changes to the delivery schedule, which prevent them from fully satisfying the customer. Moreover, the lack of transparency and effective information flow can inhibit trust and continuous improvement (Brun et al., 2020; Kotcharin et al., 2024). This also hinders root cause analysis and problem-solving initiatives. It is, therefore, critical that within the supply network, buyers and suppliers share data, align their requirements and capacities, develop demand forecasts together, agree on the technical specifications and consider quality management beyond the boundaries of the organisation and at the supply chain level.

Supplier Complacency

Supplier complacency can present a significant barrier to the development of effective buyer–supplier partnerships and SCQM programmes. Over time, suppliers may become overly comfortable with the status quo, especially if they face little competitive pressure or if the buying organisation provides limited performance feedback. This complacency can manifest in a lack of motivation to pursue quality improvements, invest in innovation, or respond proactively to the buyer's evolving needs. In some cases, long-standing relationships may inadvertently foster a sense of entitlement or reduced urgency to meet performance targets. To counteract this, buying firms must actively manage the relationship by setting clear expectations, regularly reviewing performance, and creating incentives that encourage continuous improvement and shared responsibility for quality outcomes.

It is important that the suppliers do not adopt an insular behaviour, are prepared and able to take a global and dynamic view of the supply network and are concerned with their customer satisfaction.

There are different types of measurement relating to a buyer's satisfaction with the quality of supplies, which are more or less proactive, for example:

- Quality performance and reliability analysis
- Customer assessment rating and audit reports
- Contractual requirements outlined in the customers' vendor improvement plans
- Benchmarking assessment and analysis
- Competitor evaluation
- Advanced quality planning

Misalignment or Misguided Supplier Improvement Objectives

A frequent barrier to effective supplier development and SCQI is the misalignment or misguidance of improvement objectives between buyers and suppliers. This occurs when buying organisations impose improvement targets that do not align with the supplier's capabilities, strategic direction, or operational context. For example, a buyer may prioritise quality improvement, while the supplier is focused on cost reduction or delivery reliability improvement. Such disconnects can lead to wasted resources, frustration, and a breakdown in collaboration. Moreover, improvement initiatives may be guided by the buyer's internal priorities without adequate consultation or joint planning, resulting in objectives that lack mutual ownership. To avoid this, supplier development efforts should be co-designed, with shared goals and clearly defined success criteria that reflect both parties' strategic interests and constraints.

Buyers need to be clear about what they require from their suppliers' improvement initiatives and not underestimate the time and resources needed to introduce and develop partnerships.

Finally, there also appears to be a distortion of the quality message as requirements are passed down the supply chain. For example, when faced with demands to improve quality from customers, first-tier suppliers usually react by implementing specific tools and techniques required by the buyer firm. In turn, the supplier then insists that their own suppliers use the same tools and techniques, but fails to understand that these are only fully effective within the context of an organisation-wide approach to continuous improvement.

Lack of Buyer Credibility as Viewed by Their Suppliers

Suppliers need to be convinced that the focal firm is serious about continuous improvement programmes and the SCQM initiatives. This requires the focal firm's behaviour and attitudes to be consistent with the message conveyed to the partners. The following are examples of how a credibility gap may emerge:

- Procurement management practices such as a competitive pricing policy to force down prices, frequent switches from one supplier to another, unpredictable and inflated production schedules, last-minute changes to schedules, poor engineering design/production/supplier liaison, over-stringent specifications inconsistent decisions made by the quality assurance (QA) employees, abuse of power by the QA team, and the potential use of 'loss of business' as a bargaining technique in negotiating a price reduction. It is not uncommon for a focal firm to talk about quality to its suppliers and then act quite differently by relegating quality to secondary importance behind, for example, price and meeting the production schedule.
- The quality management (QM) and business excellence image which major purchasing organisations attempt to create in discussions with suppliers is not reflected in practice when the supplier employees visit their manufacturing sites.
- The buyer fails to react positively to the supplier's concerns about design issues and is prepared to let the supplier carry the consequences.
- Failure to respond to a supplier's request for information and provide advice on queries.

Misconceptions Regarding Purchasing Power

Purchasing power is a major issue in the buyer–supplier relationship. A lack of purchasing power is a commonly cited reason for the lack of success in improving supplier performance and implementation of SCQM programmes. The general view is that a purchaser's influence on its suppliers varies with its purchasing power, and the greater this is, the more effective will be its QA activities. These power imbalances can cause uneven levels of commitment in the business relationship. However, purchasing power alone is no guarantee of improving supplier performance. Companies with considerable purchasing power may well improve the quality of purchased items, but will not necessarily achieve lasting benefits or motivate their suppliers to internalise the benefits of a process of continuous improvement to satisfy all their customers.

Conditions of Buyer–Supplier Partnership

One of the key points which stands out is the wide diversity of partnership arrangements and definitions. To get the most out of a partnership, it is vital to understand the relationship and to manage it as a 'process', its current state and how it can be developed. One of the main driving forces behind the move to buyer–supplier partnerships to enhance quality has been the establishment of Japanese transplants overseas and particularly in Europe. However, it must be recognised that the conditions under which European customers and suppliers operate are markedly different to those in Japan, where many large organisations have dedicated and even exclusive suppliers. This has led to the phenomenon, in the motor industry for example, where it is not just Toyota vying with Nissan and Honda for supremacy, but the entire Toyota supply chain battling against the Nissan and Honda supply chains. These are clearly not the conditions that operate in Europe or the USA, where dedicated suppliers are few and far between. Suppliers will work closely with a particular customer to develop a product, process or service. This includes obtaining a supplier's input on new product development (NPD) and sharing product planning and development data with suppliers (Petersen et al., 2005; Handfield & Lawson, 2007; Lawson et al., 2015). However, the benefit to that customer is likely to be short-lived because, in a commercial environment, the supplier must work with all its customers in a similar way to retain their business. There are, of course, issues of confidentiality which a first-tier supplier needs to respect when dealing with competitors involved in similar activities.

Many buyers and suppliers are abandoning adversarial relationships in favour of more cooperative partnerships. Quite rightly, different companies, industries and sectors are developing partnerships in their own way to meet their own needs and circumstances. However, it must be recognised that no one enters a partnership with their suppliers or customers out of any altruistic motive or wish to be 'nice' to them. Partnerships are driven by hard-headed business objectives, mainly the need to achieve/maintain competitiveness in an increasingly global and hostile business environment. For example, even partnership suppliers are being told that for an increasing amount of business, they are expected to cut costs and enhance quality. Therefore, it must be recognised that buyer–supplier partnerships are not an easy option or some sort of panacea. This is particularly the case where a supplier is expected to meet the global requirements of its major customers.

The Issues to be Considered in Partnership

To be effective, a partnership requires well-trained employees capable of working with suppliers to achieve the agreed objectives; in effect, these employees act as change agents. Procurement and other staff who liaise with suppliers will need to understand the capabilities of suppliers' processes, systems and value streams and have a good working knowledge of the philosophy, principles, techniques of improvement and shop-floor procedures.

It is also important that the most effective mechanisms and linkages for communication and feedback are used. Typically, purchasing, quality, design, engineering/technical and production personnel all talk to suppliers, but with no single functional area accepting total responsibility for the quality, cost and delivery of the bought-out items. The need for clear accountability and coordination is a crucial factor in ensuring that channels of communication between customers and suppliers are effective and that suppliers receive a consistent message. Importantly, it must be clear who will be responsible for all negotiations and communications for current and future business with each supplier.

It will also be the case that while a few suppliers may be world-class, the majority will need to improve if they are to meet the company's expectations. Therefore, most companies will find that they will need to adopt different practices with different suppliers. Probably, in most cases, a partnership approach based on a commitment to supplier improvement will be the order of the day. On the other hand, with those suppliers whose performance is already world-class, it may be the buyer who feels they are being improved. However, with some suppliers, relationships may well remain antagonistic.

Therefore, before starting a process of partnership, a company will need to review its supplier base and identify those suppliers with whom it needs to work in the long term and the type of relationship it will be able to establish. The strategy should reflect core organisational expertise. As it will not be possible to launch a partnership approach with all its suppliers at once, the company will need to establish a mechanism for selecting the initial group of suppliers (e.g. Kraljic matrix analysis). Assisting suppliers to improve their performance is important; however, it must be recognised that the delivery of non-conforming product from a supplier can often be attributed to an ambiguous procurement specification and poorly detailed customer requirements. The content of material and product specifications has become highly standardised, and usually includes such features as functional physical characteristics, dimensional details, reliability characteristics, methods of test and criteria for acceptance, conditions of manufacture, installation, storage and use, and so on. The procurement department should review the accuracy and completeness of these documents before they are released to suppliers. It is good practice to send these documents to the quality department for comment before transmission to the supplier.

It is also important to recognise that, just as suppliers can learn from customers, the reverse also applies. Suppliers are knowledgeable in their own field of operation and should be given every opportunity to provide design input to the preparation of the specification. With the reduction in specialist technical staff in many customer organisations, this is now a common occurrence. Suppliers will be more likely to accept responsibility for defects and their associated costs if they are involved in the design of the product or formally agree with the customer on the specification and drawing for the part to be produced. This supplier's input to the design process is a key factor in cost avoidance and reduction, and helps to reduce product development lead time.

Strategic sourcing, taking the form of single or dual sourcing, is considered by firms and leads to supplier base reduction (Cousins et al., 2008; Costantino & Pellegrino, 2010; Li, 2013). Even if some organisations have sought to avoid single sourcing and being completely dependent on a supplier to minimise the risk of disruption, it is a balance of power situation in which careful judgement is needed. A reduction in the supplier base can result in benefits such as:

- Less variation in the characteristics of the supplied product.
- Improved opportunity for improving processes, developing innovations and prototyping.
- Increases in the amount of time the customer's QA and procurement employees can devote to suppliers.
- Improved and simplified communications with suppliers.
- Less transportation, handling, and inspection activity.
- Fewer accounts to be maintained and thus reduced costs for both parties.

It is easier to develop a partnership relationship if the suppliers are near the customer (Cannon & Homburg, 2001). Consequently, several customers are now reversing their international sourcing strategies to develop shorter supply chains and are recommending that suppliers set up operations close to their main manufacturing facilities. Closeness is also a vital element in the use of a Just-in-Time (JIT) strategy (Holl et al., 2010).

Summary

Suppliers are recognised as an essential part of any firm's competitiveness, quality, and innovation performance (Lau et al., 2010; Wagner, 2010; Lawson et al., 2015). There are two major reasons for this: i) greater global specialisation and ii) changes in competition. SCQM and effective partnership require purchasing organisations to treat suppliers as long-term business partners, and this necessitates a fundamental shift from the traditional adversarial buyer–supplier relationship. A properly implemented partnership will help to reduce costs and increase market share to the benefit of both parties, together with technology transfer issues surrounding product, process, practices, and systems. However, the nature and mechanisms of partnership must be related to the circumstances and needs of those involved.

In conclusion, the following best practice will help both buyers and suppliers establish the type of partnership that is most appropriate for them:

- Look at ways of reducing the size of the supplier base. By reducing incoming material, component and sub-assembly variation and variability, outgoing product and service quality will improve.
- Ensure that, in support of the supplier development process and its various stages, the buyer organisation employees use the appropriate engineering and quality tools. These tools include statistical process control (SPC), the seven quality control tools, the seven management tools, failure mode and effects analysis (FMEA), fault tree analysis, quality function deployment (QFD), design of experiments (DoE) and benchmarking; the tools also facilitate design for manufacturability and cost avoidance.
- Involve suppliers in NPD and investigate the full range of ways of innovating.
- Award long-term contracts to key strategic suppliers who have shown commitment and improvements to demonstrate the tangible benefits that can arise from a partnership.
- Implement an assessment and rating scheme to select and measure the performance of suppliers based on multiple criteria decision analysis (MCDA). Poor selection will lead to increased costs as other suppliers are sought to compensate for the deficiencies of the one chosen without due care.
- Develop procedures, objectives, and strategies for communicating with the supply base.
- Treat suppliers as partners, thereby establishing trust, co-operation, and dependence.
- Ensure that the staff dealing with suppliers act consistently and match actions to words.
- Develop and decide upon mutually agreed-upon purposes and values that define the relationship and measure its success. The approach by the customer must be seen by the supplier as helpful, constructive and of mutual benefit.
- Decide and agree on the best means of promoting and monitoring good communication to create a constructive dialogue and ensure the provision of reliable information. This requires defined points of communication to be established.
- Provide training and education to raise awareness of the partnership approach and specific training for the new skills required.
- Ensure that buyer and supplier organisations are sufficiently knowledgeable about each other's business, products, procedures, and systems and how the respective organisations work.
- Nurture a flexible and open approach, which encourages positive constructive criticism.

References

Alkalha, Z., Dehe, B., & Reid, I. (2025). How absorptive capacity mediates supply chain effectiveness: A quality integration approach to boosting operational performance. *Operations Management Research*, 18, 870–890.

Alkalha, Z., Dehe, B., Reid, I., & Al-Zu'bi, Z.B.M. (2024). Investigating the mediating impact of supplier quality integration in pharmaceutical supply chains. *International Journal of Quality & Reliability Management*, 42(6), 1596–1624.

Alkalha, Z., Reid, I., & Dehe, B. (2019). The role of absorptive capacity within supply chain quality integration. *Supply Chain Management: An International Journal*, 24(6), 805–820.

Brun, A., Karaosman, H., & Barresi, T. (2020). Supply chain collaboration for transparency. *Sustainability*, 12(11), 4429.

Cannon, J.P., & Homburg, C. (2001). Buyer–supplier relationships and customer firm costs. *Journal of Marketing*, 65(1), 29–43.

Costantino, N., & Pellegrino, R. (2010). Choosing between single and multiple sourcing based on supplier default risk: A real options approach. *Journal of Purchasing and Supply Management*, 16(1), 27–40.

Cousins, P., Lamming, R., & Squire, B. (2008). *Strategic supply management: Principles, theories and practice*. UK: Pearson Education.

Cudney, E., & Elrod, C. (2010). Incorporating lean concepts into supply chain management. *International Journal of Six Sigma and Competitive Advantage*, 6(1–2), 12–30.

Dale, B., Bamford, D., & van der Wiele, T., (2016). *Managing quality: An essential guide and resource gateway*, 6th edition (p. 330). UK: John Wiley & Sons.

Fawcett, S.E., Jones, S.L., & Fawcett, A.M. (2012). Supply chain trust: The catalyst for collaborative innovation. *Business Horizons*, 55(2), 163–178.

Fernandes, A.C., Sampaio, P., Sameiro, M., & Truong, H.Q. (2017). Supply chain management and quality management integration: A conceptual model proposal. *International Journal of Quality & Reliability Management*, 34(1), 53–67.

Handfield, R.B., & Lawson, B. (2007). Integrating suppliers into new product development. *Research-Technology Management*, 50(5), 44–51.

Heinis, S., Bamford, D., Papalexi, M., & Vafadar, A. (2022). Services procurement: A systematic literature review of practices and challenges, *International Journal of Management Reviews*, 24(3), 352–372.

Holl, A., Pardo, R., & Rama, R. (2010). Just-in-time manufacturing systems, subcontracting and geographic proximity. *Regional Studies*, 44(5), 519–533.

Kim, H.J., Son, J., & Kim, S.W. (2016). Strategy for improving efficiency of supply chain quality management in buyer-supplier dyads: The suppliers' perspective. *Mathematical Problems in Engineering*, 2016(1), 8641702.

Kotcharin, S., Dehe, B., & Boonchoo, P. (2024). 'All you need is trust': Examining trust, information sharing, and supplier performance perception in SMEs from a social exchange perspective. *Cogent Business & Management*, 11(1), 2415523.

Lau, A.K., Tang, E., & Yam, R.C. (2010). Effects of supplier and customer integration on product innovation and performance: Empirical evidence in Hong Kong manufacturers. *Journal of Product Innovation Management*, 27(5), 761–777.

Lawson, B., Krause, D., & Potter, A. (2015). Improving supplier new product development performance: The role of supplier development. *Journal of Product Innovation Management*, 32(5), 777–792.

Li, C. (2013). Sourcing for supplier effort and competition: Design of the supply base and pricing mechanism. *Management Science*, 59(6), 1389–1406.

Luo, L., Liu, X., Zhao, X., & Flynn, B.B. (2023). The impact of supply chain quality leadership on supply chain quality integration and quality performance. *Supply Chain Management: An International Journal*, 28(3), 508–521.

Minshull, L.K., Dehe, B., & Kotcharin, S. (2022). Exploring the impact of a sequential lean implementation within a micro-firm–a socio-technical perspective. *Journal of Business Research*, 151, 156–169.

Petersen, K.J., Handfield, R.B., & Ragatz, G.L. (2005). Supplier integration into new product development: Coordinating product, process and supply chain design. *Journal of Operations Management*, 23(3–4), 371–388.

Siagian, H., Tarigan, Z.J.H., & Jie, F. (2021). Supply chain integration enables resilience, flexibility, and innovation to improve business performance in COVID-19 era. *Sustainability*, 13(9), 4669.

So, S., & Sun, H. (2010). Supplier integration strategy for lean manufacturing adoption in electronic-enabled supply chains. *Supply Chain Management: An International Journal*, 15(6), 474–487.

Touboulic, A., & Walker, H. (2015). Love me, love me not: A nuanced view on collaboration in sustainable supply chains. *Journal of Purchasing and Supply Management*, 21(3), 178–191.
Wagner, S.M. (2010). Supplier traits for better customer firm innovation performance. *Industrial Marketing Management*, 39(7), 1139–1149.
Yu, Y., & Huo, B. (2018). Supply chain quality integration: Relational antecedents and operational consequences. *Supply Chain Management: An International Journal*, 23(3), 188–206.
Zhang, M., & Huo, B. (2013). The impact of dependence and trust on supply chain integration. *International Journal of Physical Distribution & Logistics Management*, 43(7), 544–563.
Zimon, D., Urbaniak, M., Madzík, P., & Prokopiuk, I. (2022). Supply chain quality management (SCQM) literature Review and model proposal in the Era Of Industry 4.0. *International Journal for Quality Research*, 16(4), 1283–1296.

Part 3

Quality Management Tools, Techniques, and Systems

7 Quality Management Tools

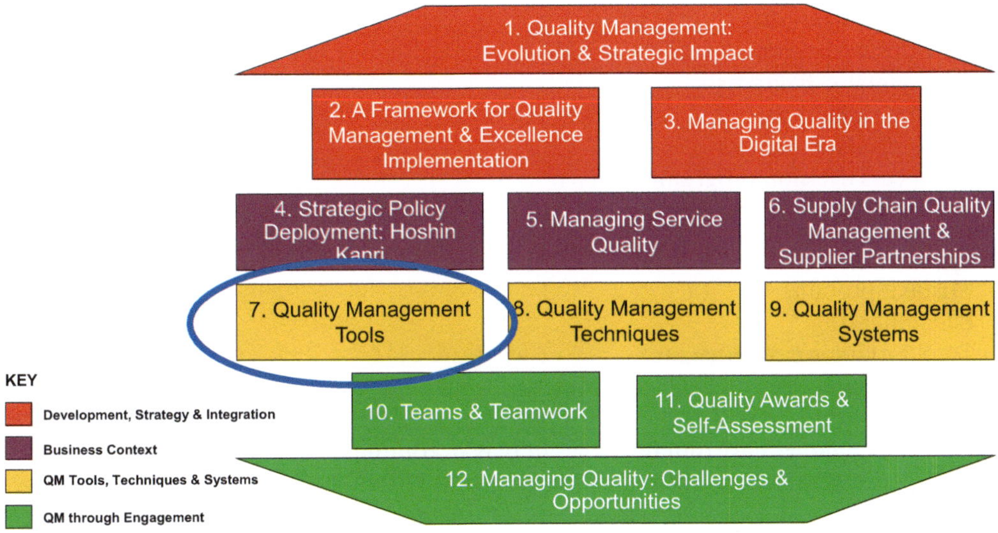

Introduction

In this chapter we provide an overview of the tools which are likely to be used in an organisation's improvement process. The focus is on describing the tools and their uses, and avoids detailed discussion of how to construct a Process Decision Programme Chart, for example. A deliberate attempt has been made to choose examples from a variety of industries to give the reader a flavour of their applicability in a wide number of situations.

To support, develop and advance a process of continuous improvement, it is necessary for an organisation to use a selection of tools and techniques. Some of these tools and techniques are simple, while others are more complex. There are a considerable number of tools and techniques; the following are perhaps the most popular and best known:

- Checklists
- Flowcharts
- The seven quality control tools:
  - cause-and-effect diagram
  - check sheet

DOI: 10.4324/9781003594369-10

  - control chart
  - graphs
  - histogram
  - Pareto diagram
  - scatter diagram
- Statistical process control (Six Sigma)
- Failure mode and effects analysis
- Fault tree analysis
- Design of experiments
- Quality function deployment
- The seven management tools:
  - affinity diagrams
  - relations diagrams
  - systematic diagrams
  - matrix diagrams
  - matrix data analysis
  - process decision programme chart
  - arrow diagrams
- Departmental purpose analysis
- Mistake-proofing
- Benchmarking
- Total productive maintenance
- Housekeeping & 5s

Tools and techniques have different roles to play in continuous improvement, and if applied correctly, give repeatable and reliable results (Bamford & Greatbanks, 2005). Their roles include:

- Summarising data and organising its presentation
- Data-collection and structuring ideas
- Identifying relationships
- Discovering and understanding a problem
- Implementing actions
- Finding and removing the causes of the problem
- Selecting problems for improvement and assisting with the setting of priorities
- Monitoring and maintaining control
- Planning
- Performance measurement and capability assessment

Several of the tools and techniques in the above list will be detailed in the following chapters.

Selecting Tools and Techniques

The potential user must always be aware of the main uses of the tool or technique they are considering applying. There is often a danger of using a tool or technique in a blinkered manner, almost expecting it to solve the problem automatically. For example, the seven quality control tools mentioned above are designed to aid data analysis, process control and quality improvement initiatives, while the Seven Quality Management Tools were developed to visually display data (McDermott et al., 2023).

When selecting tools and techniques, there are two factors that organisations should keep in mind:

- The application of any tool or technique in isolation without a strategy and plan will only provide short-term benefits. If tools and techniques are to be effective over the longer term, appropriate employee behaviour and attitudes are needed to make effective improvements.
- No one tool or technique is more important than another – they all have a role to play at some point in the improvement process. It is a mistake to single out one tool or technique for special attention and to become over-reliant on it. Kaplan's 'Law of the Instrument' tells us that if you only have a hammer, it is surprising how many problems look like nails!

Several companies have used tools and techniques without thinking through the implications of their use. This can create misconceptions and misunderstandings, which can become barriers to progress. Some companies single out a specific tool or technique, sometimes seemingly at random, and apply it with undue haste without giving sufficient thought to the following issues:

- What is the fundamental purpose of the technique?
- What will it achieve?
- Will it produce benefits if applied on its own?
- Is the technique right for the company's product, processes, people, and culture?
- How will the technique facilitate improvement?
- How will it fit in with, complement or support other techniques, methods and quality management systems already in place, and any that might be introduced in the future?
- What organisational changes, if any, are necessary to make the most effective use of the technique?
- What is the best method of introducing and then using the technique?
- What are the resources, skills, data, information, training, etc. required to introduce the technique successfully?
- Does the company have the management skills and resources and the commitment to make the technique work successfully?
- What are the potential difficulties in using the technique?
- What are the limitations, if any, of the technique?

It is important for managers to address these questions when considering the introduction of any tool or technique. Unfortunately, some managers try to use these tools and techniques as a quick-fix solution to the problems facing their organisation at a particular point in time. In general, management teams that are 'technique reactive' tend to be unclear on the concept of QM and strategic process improvement. They often confuse the implementation of a particular technique with QM and tend to use the technique as an end in itself rather than as a means to an end.

If the management team is preoccupied with specific techniques and lacks an adequate understanding of QM and the improvement process, the risk is that tools and techniques are picked up and discarded as fashion changes. When this happens and a tool or technique fails to meet expectations, disillusionment sets in, and the company experiences 'change fatigue' – difficulty convincing its employees that it is serious about improvement, which has an adverse effect on the use of techniques in the organisation in the future.

One of the main reasons that companies fall into this trap is that they have unduly high expectations of the benefits arising from the use of a single tool or technique, which stems from a lack of clarity and in-depth understanding of it. This can be a result of the publicity and marketing which accompany the launch of a specific tool or technique. A single tool or technique may simply indicate or signify the presence of a problem, which, when resolved, can produce only a small incremental improvement. It is only because of the systematic application of a series of tools and techniques within a QMS that the organisation will see long-term benefits from its improvement initiatives (see Figure 7.1).

Motivation for the use of any particular tool or technique is a key factor in the success of its implementation (Bamford & Greatbanks, 2005). They could be those specified as a contractual requirement by a major customer, they may be what management believes the marketplace will be expecting in the future, or the view may be taken that their use will give the organisation an edge over its competitors.

Because of the variety of starting points and motivations for improvement, it is not possible to identify a universal implementation plan detailing the order in which specific techniques should be used by an organisation. However, one piece of advice is that organisations should start with the simpler techniques, such as checklists, flowcharts and the seven original quality control tools. Simple tools and techniques can be just as effective as the more complex ones. Simple tools also have the advantage that there are IT packages that are readily and freely available to allow their implementation; the more complex tools often require bespoke software packages or training before they can be implemented successfully.

Difficulties and Issues Relating to the Use of Tools and Techniques

Research carried out by Mathur et al. (2023) into the use of QM tools in Higher Education, discovered that the critical success factors relating to the successful use and application of tools and techniques could be grouped into the following categories:

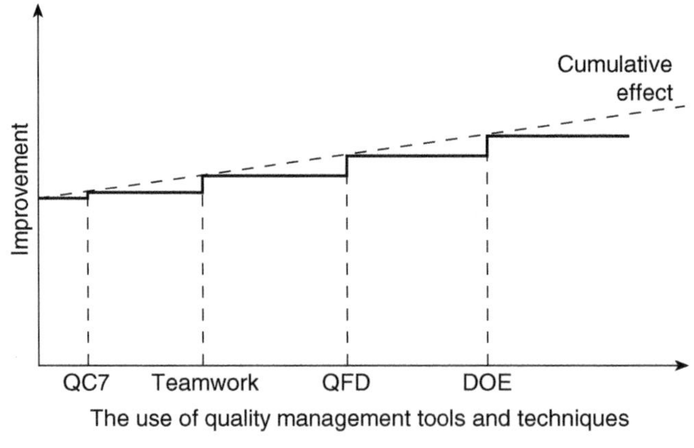

Figure 7.1 Incremental improvement through the use of quality management tools and techniques.

- Management support
- Widespread training
- Having a continuous improvement programme
- Communication about the need for the tools from senior management
- Providing the required resources for improvement projects
- Having the opportunity to use the tools
- Opportunity to be involved in problem solving
- Sharing success stories and benefits

The authors acknowledge that these factors would differ in different industries, for example, training in QM tools is much more widespread in the service and manufacturing sectors, and that service organisations tend to avoid the tools that require more in-depth quantitative analysis.

Problem-Solving Methodology

The use of tools and techniques should be employed within a problem-solving approach for maximum effectiveness and efficiency. Probably the best-known problem-solving cycle is the Plan, Do, Check, Act (PDCA) Cycle developed by W. Edwards Deming and shown in Figure 7.2.

The *plan* aspect of the cycle is usually considered in four stages:

1 Define the problem or improvement opportunity and specify objectives.
2 Identify the likely causes of the problem.
3 Pinpoint the root causes of the problem.
4 Prepare solutions and develop and agree on an action plan.

The *do* is concerned with implementing the action plan; *check* monitors the effectiveness of the actions which have been implemented, and *act* relates to standardisation of the results and transferring the practices to other processes. At each stage of the cycle, a range of tools and techniques is employed. The Ford Motor Company, amongst other organisations, developed a step-by-step process of ensuring that any improvement action is permanent.

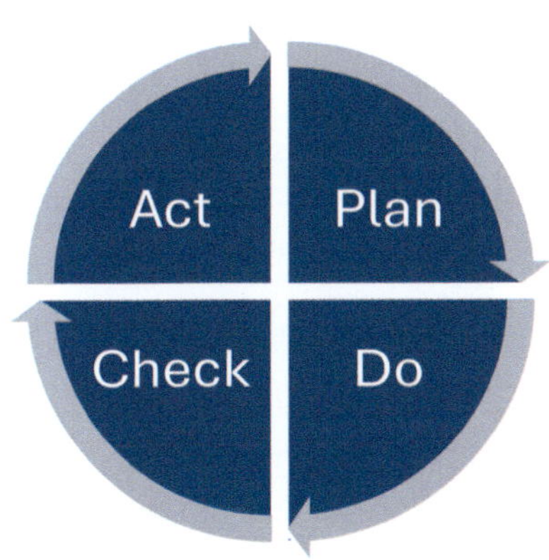

Figure 7.2 The plan, do, check, act cycle.

This is known as the TOPS (Team-Oriented Problem-Solving) 8D (eight-discipline) approach.

The following is a summary of the eight disciplines:

- D1 Use a team approach
- D2 Describe the problem:
  - Review and analyse existing data
  - Establish problem definition and statement
  - Develop the problem profile
  - Confirm the problem with the customer
- D3 Implement and verify interim containment actions (ICA):
  - Choose the best ICA
  - Test for feasibility
  - Develop an action plan and implement ICA to isolate the problem from the customer until permanent corrective action is available
  - Monitor and report the effectiveness of ICA
- D4 Define and verify root cause(s):
  - Review, improve and update the problem definition and description
  - Identify possible causes by comparison to the problem description
  - Select likely causes
  - Verify root causes
  - Report
- D5 Choose and verify permanent corrective actions (PCA):
  - Choose the best PCA
  - Re-evaluate ICA
  - Verify PCA
  - Make a choice
  - Report
- D6 Implement permanent corrective action:
  - Develop implementation plan
  - Remove ICA
  - Implement PCA
  - Monitor the process and assess the effectiveness of problem elimination
  - Formalise changes, and update documents and the system
  - Notify affected personnel
  - Report
- D7 Prevent recurrence:
  - Review current process
  - Identify critical areas or supporting PCA
  - Make recommendations for improvement of current processes
  - Report
- D8 Congratulate the team

It may be useful for you to carry out a review of the application and implementation of the tools and techniques by your organisation. Table 7.1a provides a checklist to allow you to assess not only those tools which are known about and compare these to those tools that are actually used, but also to assess each of the tools in turn.

Table 7.1a An assessment grid for a health check: recognition and use grid

Example	*Recognise*	*Use*
QC7		
Cause and effect	√	
Check sheets		
Control charts		
Graphs/charts	√	√
Histograms	√	√
Pareto analysis	√	√
Scatter diagrams	√	
QC7		
Cause and effect		
Check sheets		
Control charts		
Graphs/charts		
Histograms		
Pareto analysis		
Scatter diagrams		
M7		
Affinity diagrams		
Arrow diagrams/critical path analysis		
Matrix data analysis methods		
Matrix diagrams		
Process decision programme chart		
Relation diagrams		
Systematic diagrams/tree diagrams		
Techniques		
Benchmarking		
Brainstorming/brainwriting		
Departmental purpose analysis		
Design of experiments (Taguchi)		
Failure mode and effects analysis		
Flow charts		
Force field analysis		
Problem-solving methodology		
Quality costs		
Quality function deployment		
Questionnaire		
Sampling		
Statistical process control		
*		
*		

Please indicate on the grid ONLY the techniques and tools you recognise. For those that you have recognised, if you use them for ***any*** purpose, not only for quality-related matters, please tick the box marked use.

* Add any company-specific techniques and tools not indicated on the list.

Table 7.1b An assessment grid for a health check: application grid

	Importance	*Relevance*	*Use*	*Understand*	*Application*	*Resources*	*Management*	*Training*	*Benefit*
For example: *Pareto analysis*	5	4	2	3	3	2	1	1	4
QC7									
Cause and effect									
Check sheets									
Control charts									
Graphs/charts									
Histograms									
Pareto analysis									
Scatter diagrams									
QC7									
Affinity diagrams									
Arrow diagrams/critical path analysis									
Matrix data analysis methods									
Matrix diagrams									
Process decision programme chart									
Relation diagrams									
Systematic diagrams/tree diagrams									
Techniques									
Benchmarking									
Brainstorming/brainwriting									
Departmental purpose analysis									
Design of experiments (Taguchi)									
Failure mode and effects analysis									
Flow charts									
Force field analysis									
Problem-solving methodology									
Quality costs									
Quality function deployment									

Questionnaire
Sampling
Statistical process control (Six Sigma)
Other techniques, tools, systems
For example:
ISO9000 series
Quality operating system QS9000
Other awards (e.g. EQA)

Please complete the grid ONLY for the techniques or tools you indicated on the recognition and use grid.
Do not attempt to fill it in its entirety. They may be occasions when some of the categories cannot be allocated a score; in that case insert 9 (not applicable)Score out of 5 in each of the categories where:
1 = No value
2 = Low value (e.g. little used, not understood, little or poor training, etc.)
3 = Some value (e.g. basic understanding, small benefits, basic training, etc.)
4 = High value (e.g. good understanding, some benefits, reasonable training, etc.)
5 = Very high value (e.g. complete understanding, excellent benefits, effective training, etc.)

Checklists

Checklists (sometimes called inspection or validation checklists) are used to highlight the key features of a process, equipment, system, and/or product/service to which attention needs to be given, and ensure that the procedures for an operation, housekeeping, inspection, and maintenance have been followed. Checklists are also used in audits of both products and systems. While they are a very simple tool, they are an invaluable aid for quality assurance.

The basic steps in constructing a checklist are:

- Study the activity for which the checklist is to be drawn up.
- Drawing on the observations of the process, discussions with operatives, and appropriate working instructions and procedures, construct the checklist.
- Walk the checklist through the process by following what happens at each stage.
- Ask the person who is carrying out the process to check its accuracy.
- Display it next to the process.
- Assess its use in practice.

Table 7.1b is an example of a checklist to identify the use of Quality Tools in your organisation.

Flowcharts

Process mapping (sometimes called 'blueprinting' or process modelling) in either a structured or unstructured format, is required in order to develop an in-depth understanding of a process, before the application of quality management tools and techniques such as Failure Mode and Effects Analysis (FMEA), Statistical Process Control (SPC) and Six Sigma. A flowchart is employed to provide a diagrammatic picture, often by means of a set of established symbols, showing all the steps or stages in a process, project, or sequence of events and is useful to document and describe a process as an aid to understanding, examination, and improvement.

A flowchart, when used in a manufacturing context, may show the complete process from goods receipt through storage, manufacture, and assembly to dispatch of the final product, or simply some part of this process in detail. This type of process map focuses on the flow of materials; other process maps may detail the flow of information or people through a process (Bamford, Forrester, & Reid, 2023). What is important is that each 'activity' is included, to focus attention on aspects of the process or subset of the process where problems have occurred or may occur, to enable some corrective action to be taken or countermeasures put into place.

Traditionally, flowcharts (called process maps) have employed conventional symbols to define activities such as operation, inspection, delay or temporary storage, permanent storage and transportation, and are widely used by operations managers and industrial engineering personnel. In more recent times they have witnessed considerable use in business process re-engineering and service improvement efforts.

There are several variants of the classical process flowchart, including those tailored to an individual industry or company use, with different symbols being used to reflect the situation under study. What is important is not the format of the chart and/or flow diagram, but that the process has been mapped out with key inputs, value-adding steps and outputs defined, and that it is understood by those directly involved and responsible for initiating improvements. Analysing the data collected on a flowchart can help to uncover irregularities and potential problem points. Also, in some organisations people are only aware of their own aspect of a process and process mapping helps to facilitate a greater understanding of the whole process: it is essential to the development of the internal customer–supplier relationship. Figure 7.3 is an example of flowcharting for the process of non-conformance identification and preventative action.

There are a number of different flowcharting methodologies, for example, Value Stream Mapping, which differ in the symbols and data used to create the chart; however, the following are the basic steps in constructing a flowchart:

- Define the process and its boundaries, including start and endpoints.
- Decide the type and method of charting and the symbols to be used, and do not deviate from the convention chosen.
- Decide the level of detail with which the process is to be mapped.
- Describe the activities, in sequence, in the process using the agreed methodology.
- Assess if these stages are in the correct sequence.
- Ask people involved with the process to check its veracity.

Checksheets

These are a sheet or form used to record data. The checksheet is a simple and convenient recording method for collecting and determining the occurrence of events.

Checksheets are helpful in following the maxim of 'no checking or measurement without recording the data' and are useful for an initial analysis of a problem.

They are prepared, in advance of the recording of data, by the operatives and staff affected by a problem. Checksheets, in table, process, diagram, or picture format, are extremely useful as a data-collection device and a record to supplement quality control charts. The data from a checksheet provides the factual basis for subsequent analysis and corrective action. There are many kinds of checksheets: Figure 7.4 gives an example of a spreadsheet-based checksheet used to record the reasons why hospital operating lists start late.

The following are the main steps in constructing a checksheet:

- Decide the type of data to be illustrated. The data can relate to the number of defectives, percentage of total defectives, cost of defectives, type of defective, process, equipment, shift, business unit, operator, etc.
- Decide which features/characteristics and items are to be checked.
- Determine the type of checksheet to use (i.e. tabular form or defect position chart).
- Design the sheet; ideally, it should be flexible enough to allow the data to be arranged in a variety of ways. Data should always be arranged in the most meaningful way to make the best use of it.
- Specify the format, instructions, and sampling method for recording the data, including the use of appropriate symbols.
- Decide the time period over which data are to be collected.

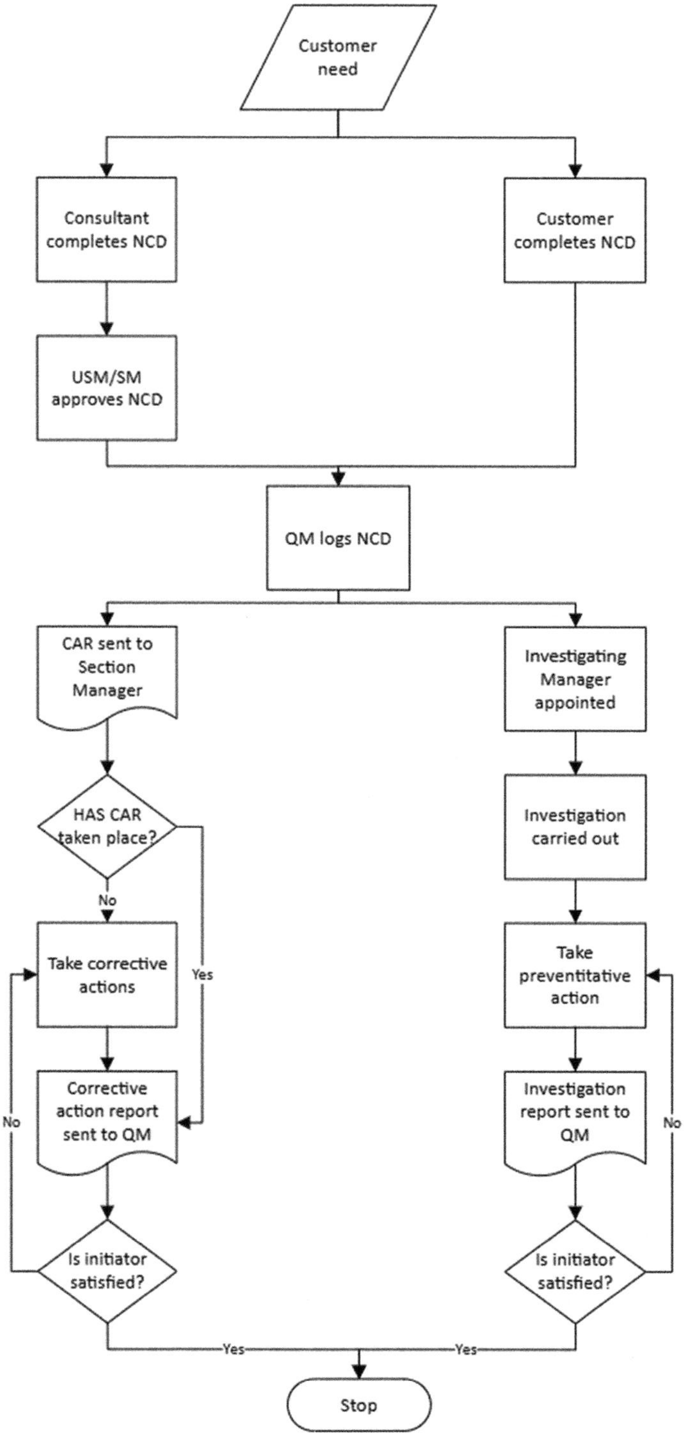

Figure 7.3 Flowchart: non-conformance identification and preventative action process.

Key: CA = corrective action; CAR = corrective action request; NCD = non-conformance document; QM = quality manager; SM = service manager; USM = unit sales manager

Late starts by reason code

Reason	Theatre										
	1	2	3	4	5	6	7	8	9	10	Total
Anaesthetist delayed	0	2	0	0	3	2	2	1	2	0	**12**
Surgeon delayed	2	2	0	0	1	8	4	3	4	7	**31**
First Operation cancelled	0	0	0	0	0	0	0	0	0	0	**0**
Previous List overran	3	0	0	0	0	1	2	0	1	3	**10**
Agreed late Start	0	0	0	1	0	0	0	0	1	0	**2**
No Beds	1	0	0	0	0	3	0	0	2	2	**8**
Patient not ready/Unwell/Not consented	2	0	0	0	2	1	0	0	1	1	**7**
Order of list changed	1	0	0	0	1	1	0	0	0	1	**4**
Urgency	1	2	0	0	0	0	1	0	0	0	**4**
No instrument/Equipment failure	0	0	0	0	0	0	0	0	0	0	**0**
Other	11	15	1	1	3	1	2	2	1	0	**37**
Total	**21**	**21**	**1**	**2**	**10**	**17**	**11**	**6**	**12**	**14**	**115**

Figure 7.4 Checksheet: for operating theatre list late starts.

Tally Charts and Histograms

Tally charts are a descriptive presentation of data and help to identify patterns in the data. They may be used as checksheets with attribute data (pass/fail, present/absent), but are more commonly used with measured or variable data (e.g. temperature, weight, and length) to establish the pattern of variation displayed, prior to the assessment of capability and computation of process capability indices (see SPC for details). Tally charts are regarded as simple or crude frequency distribution curves and provide a quick way of recording and displaying data.

Statisticians tend to construct histograms rather than tally charts, but for general analysis purposes, they are the same. A histogram is a graphical representation of individual measured values in a data set according to the frequency or relative frequency of occurrence. It takes measured data from the tally sheet and displays its distribution using the class intervals or value as a base – it resembles a bar chart with the bars representing the frequency of data over a range of values. The histogram helps to visualise the distribution of data and, in this way, reveals the amount of variation within a process, and/or other factors such as edited data and poor sampling techniques. It can be used to assess performance to a given standard, specification or tolerance. There are several forms which should be recognised – normal, skewed, bimodal, isolated island, etc. There are several theoretical models which provide patterns and working tools for various shapes of distribution. The shape of distribution most often encountered is called Normal or Gaussian.

Figure 7.5 shows a tally chart; Figure 7.6 is a histogram which has been constructed from the data collected on the tally chart.

There are several ways of constructing histograms depending upon whether the data are discrete or continuous, whether they are single or grouped values, and whether there is a vast amount of data or not. The following guidelines are given for the treatment of continuous data of sufficient quantity, for which grouping is required.

- Subtract the smallest individual value from the largest.
- Divide this range by 8 or 9 to give that many classes or groups.
- The resultant value indicates the width or interval of the group. This should be rounded for convenience, for example, 4.3 could be regarded as either 4 or 5 depending upon the data collected.

Time (seconds)	Frequency	Total
0 - 29		0
30 - 59	I	1
60 - 89	IIII	4
90 - 119	𝍸 III	8
120 - 149	𝍸 𝍸 III	13
150 - 179	𝍸 𝍸 IIII	14
180 - 209	𝍸 𝍸 II	12
210 - 269	IIII	4
170 - 299	III	3
300 - 329	III	3
330 - 359	I	1

Figure 7.5 Tally chart: Process timings.

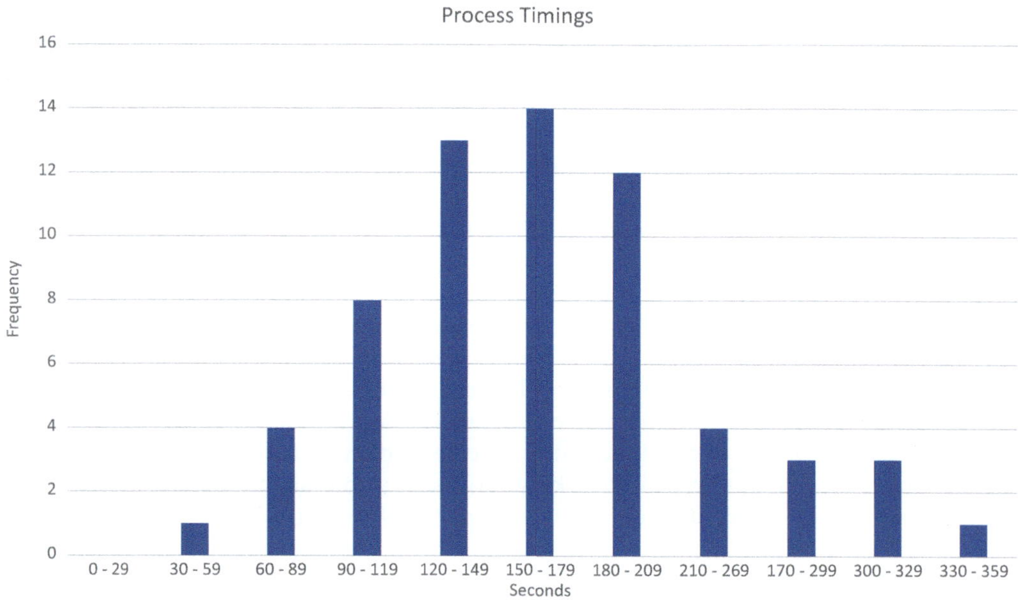

Figure 7.6 Histogram: Process timings.

- These minor calculations are undertaken to give approximately eight or nine group class intervals of a rational width.
- Each individual measurement now goes into its respective group or class.
- Construct the histogram with measurements on the horizontal scale and frequency (or number of measurements) on the vertical scale.

- The 'blocks' of the histogram should adjoin each other, that is, there should be no gaps unless there is a recorded zero frequency.
- Clearly label the histogram and state the source of the data.

Graphs

Graphs, be they presentational (i.e. to convey data in a pictorial way), or mathematical (i.e. into which data may be interpolated or from which they may be extrapolated), are used to facilitate understanding and analysis of the collected data, investigate relationships between factors, attract attention, indicate trends and make the data memorable (a picture is worth a thousand words).

There is a wide choice of graphical methods available (line graphs, bar charts, pie charts, Gantt charts, radar charts, band charts) for different types of applications. Figure 7.7 is a line graph which illustrates right-first-time production; Figure 7.8 is a bar chart showing right-first-time production.

The following are the types of issues which need to be considered in the construction of graphs:

- Use clear titles and indicate when and how the data were collected (i.e. the theme of the graphs and the source of data).
- Ensure that the scales are clear, understandable, and represent the data accurately.
- When possible, use symbols for extra data to provide clarity of explanation.
- Always keep in mind the reason why a graph is being used (i.e. to highlight some information or data in a striking and unambiguous way); anything which facilitates this objective is desirable.

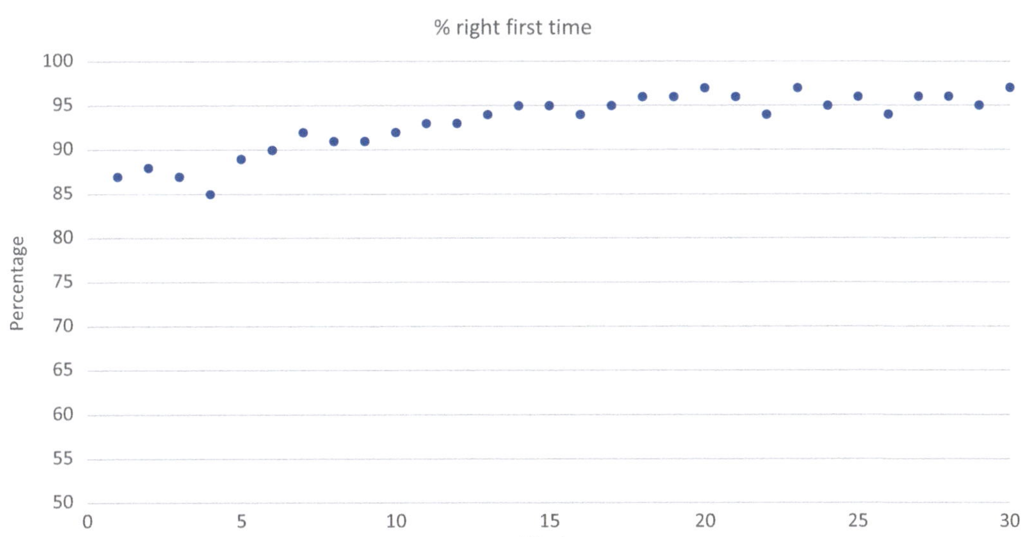

Figure 7.7 Line graph: Right-first-time operation.

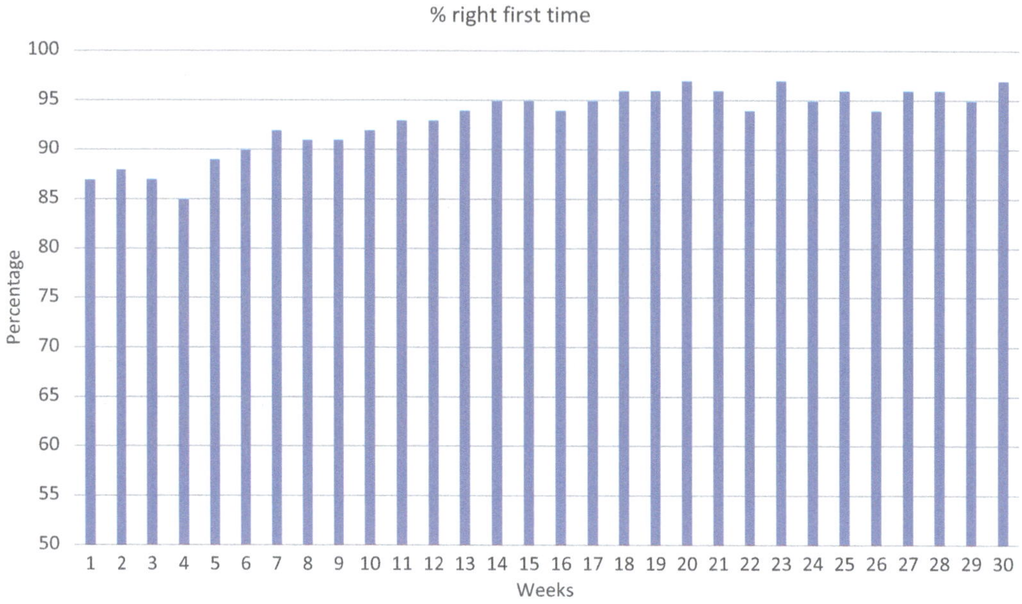

Figure 7.8 Bar chart: Right-first-time operation.

Pareto Analysis

This is a technique employed for prioritising problems of any type, for example, quality, production, complaints, stock control, sickness, absenteeism, accident occurrences, and resource allocation. The analysis highlights the fact that most problems come from a few causes, and it indicates what problems to solve and in what order (e.g. Juran's (1988) 'vital few and trivial many'). In this way improvement efforts and resources are directed where they will have the greatest impact. Pareto analysis is an extremely useful tool: it can be used to compare before and after situations and data over different time periods. It can provide insights, problem-solving, and process improvement. A Pareto diagram can be considered as a special form of bar chart, comprising a simple bar chart with a cumulative percentage curve overlaid on it.

The diagram is named after a nineteenth-century Italian economist, Vilfredo Pareto, who observed that a large proportion of a country's wealth is held by a small proportion of the population (hence the expression 'the 80/20 rule'). In the 1950s Juran, using a similar analogy, observed that a large proportion of quality problems were attributable to a small number of causes (i.e. 80 per cent of rejections are caused by 20 per cent of defect types).

The technique involves ranking the collected data, usually via a checksheet, with the most commonly occurring problem at the top and the least common at the bottom. The contribution of each problem to the grand total is expressed as a percentage, and cumulative percentages are used in compounding the effect of these problems. The ranking of the problems is usually in terms of occurrence and/or cost – just because one defect type happens more frequently than another, it does not necessarily mean that it is the costliest or the one that should be tackled first. The results are often presented in two ways: (1) ranked data as a bar chart and (2) cumulative percentages as a graph. Figure 7.9 is an analysis of the reasons for returned goods.

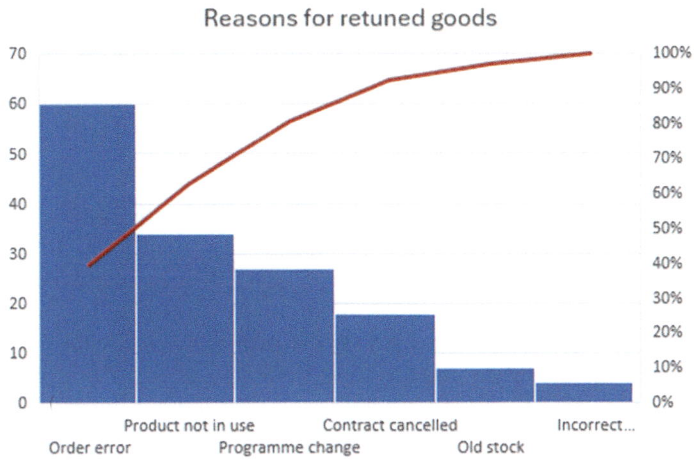

Figure 7.9 Pareto analysis: Reasons for returned goods.

Pareto analysis can be applied to a number of circumstances, for example, 80% of value comes from just 20% of available volume / assets / capacity (Bamford et al., 2023)

Pareto analysis, while simple in terms of its construction, is extremely powerful in presenting data by focusing attention on the major contributor(s) to a quality problem in order to generate attention, efforts, ideas and suggestions to hopefully gain a significant overall reduction in these problems.

The following are the basic steps in constructing a Pareto diagram:

- Agree on the problem which is to be analysed.
- Decide the time period over which data are to be collected.
- Identify the main causes or categories of the problem.
- Decide how the data will be measured.
- Collect the data using, for example, a checksheet.
- Tabulate the frequency of each category and list in descending order of frequency (if there are too many categories, it is permissible to group some into a miscellaneous category, for the purpose of analysis and presentation).
- Arrange the data as a bar chart.
- Construct the Pareto diagram with the columns arranged in order of descending frequency.
- Determine cumulative totals and percentages and construct the cumulative percentage curve, superimposing it on the bar chart.
- Interpret the data portrayed on the diagram.

Cause-and-Effect Diagrams

This type of diagram was developed by Ishikawa (1976) to determine and break down the main causes of a given problem. Cause-and-effect diagrams are also called Ishikawa diagrams, or 'fishbone' diagrams, because of their appearance. They are usually employed where there is only one problem and many possible causes.

The effect (a specific problem or a quality characteristic/condition) is the head, and potential causes and sub-causes of the problem or quality characteristic/condition are the bone structure of the fish. The diagrams illustrate in a clear manner the possible relationships between some identified effects and the causes influencing them. They also assist in uncovering the root causes of a problem and in generating ideas for improvement.

They are typically used as part of a brainstorming exercise to solicit ideas and opinions as to the possible major cause(s) of a problem, and subsequently to develop recommendations to resolve or counteract it.

It is important to define the problem or issue clearly, giving as much detail as possible to enable the identification of potential causes. This can be quite a difficult task; the problem must be manageable by the group, and it may be necessary to divide the problem into a number of smaller issues to ensure that the team's efforts and contributions are maximised.

There are several standard ways to create a cause-and-effect diagram, which generally consist of using different labels for the main branches and having different numbers of branches. In manufacturing 4Ms (machinery, manpower, method, material), 5Ms (machinery, manpower, method, material, maintenance), or 6Ms (machinery, manpower, method, material, maintenance, Mother nature) are commonly used. In non-manufacturing areas the 4Ps (policies, procedures, people, plant) are sometimes found to be more appropriate.

The exact format of the diagram is not as important as the process of bringing about appropriate countermeasures for the identified and agreed major cause(s) of the problem.

Figure 7.10 shows an example of a cause-and-effect diagram for a non-value-added work analysis carried out by a purchasing department.

Cause-and-effect diagrams are usually produced via a team approach and involve the following basic steps:

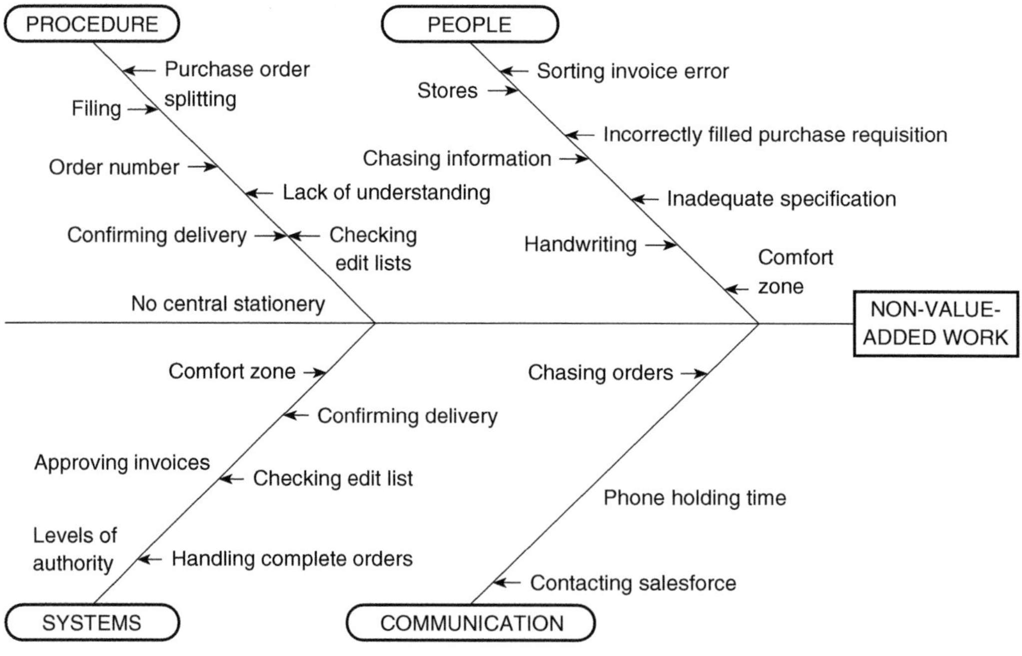

Figure 7.10 Cause-and-effect analysis: Purchasing department non-value-added work analysis.

- Define with clarity and write in a box to the right the key symptom or effect of the problem, and draw a horizontal line from the left of the box.
- Ensure that every team member understands the problem and collectively develops a clear problem statement.
- Decide the major groupings or categories for the causes of the effect; these form the main branches of the diagram.
- In a brainstorming session the group members speculate on causes of the effect, and these are added to the branches or sub-branches of the diagram.
- In a following session the causes are discussed and analysed to determine those which are most likely to have caused the effect.
- The most likely, or major, causes of the problems are ranked, by the group, in order of importance. This can be done by Pareto voting: 80 per cent of the votes should be cast for 20 per cent of the causes. (If, for example, there are 35 causes, using the figure of 20 per cent gives each member seven votes to allocate to what they believe are the causes of the effect.)
- Additional data are sometimes gathered to confirm the key causes.
- Improvement plans, actions, tests, and experiments are decided upon to both verify and address the key causes.

Scatter Diagrams and Regression Analysis

Scatter diagrams or scatter plots are used when examining the possible relationship or association between two variables, characteristics or factors; they indicate the relationship as a pattern – cause and effect. For example, one variable may be a process parameter (e.g. temperature, pressure, screw speed), and the other may be some measurable characteristic or feature of the product (e.g. length, weight, thickness). As the process parameter is changed (independent variable), it is noted, together with any measured change in the product variable (dependent variable), and this is repeated until sufficient data has been collected. The results, when plotted on a graph, will give what is called a scatter graph, scatter plot or scatter diagram. In very simple terms, variables that are associated may show a linear pattern and those that are unrelated may portray a non-linear random pattern. An example of a scatter plot showing a strong positive correlation between the amount spent on advertising and the number of customer enquiries is given in Figure 7.11.

Analysis should concern itself with the dispersion of the plots, and if a linear, or known non-linear, relationship exists between the two variables. In this way the scatter diagram is a valuable tool for diagnosis and problem-solving. Regression analysis would subsequently be used not only to establish lines of 'best fit', but to provide the basis for making estimates or predictions of the product variable for a given value of the process parameter. In this way, it is possible to reduce the amount of data which is measured, collected, plotted and analysed.

How valid or reliable such estimates are is largely a function of the degree of correlation which exists between the two variables (if indeed only two variables are under consideration) and whether the estimates are interpolated (i.e. from within the range of collected data) or extrapolated (i.e. outside that range).

Where there are more than two variables, multivariate regression analysis should be used, but a good background of statistical knowledge is required to undertake this type of analysis.

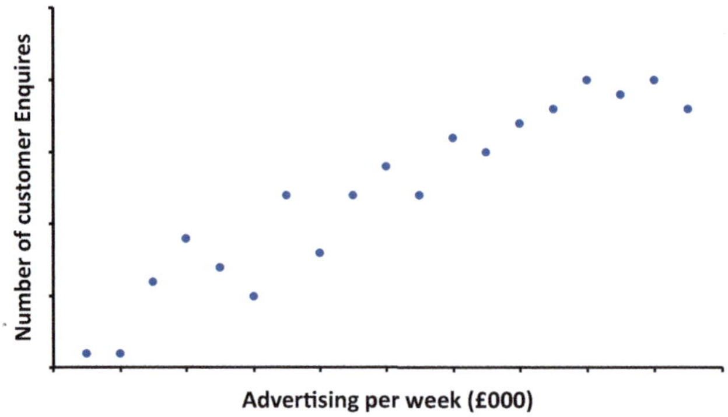

Figure 7.11 Scatter plot: Customer enquiries vs Advertising expenditure.

The Seven Management Tools

The 'seven new management tools' of quality control have now been in use for some time, having been described in detail by Mizuno (1988), Ozeki and Asaka (1990) and Barker (1989). They were developed to collect and analyse non-qualitative and verbal data from sales and marketing and design and development activities. Most of the tools have been previously used in other areas such as value engineering and value analysis, critical path analysis, programme evaluation and review technique (PERT), and organisational analysis. The tools are typically used by quality teams in sales and design areas, and in quality function deployment. It is usual to find some of the tools used together (e.g. a systematic diagram being produced from the data contained in an affinity diagram). The tools are discussed briefly below.

Relationship Diagrams

This is used to identify, understand and clarify complex cause-and-effect relationships to find the causes of and solutions to a problem and to determine the key factors in the situation under study. They are also employed to identify the key issues for a desired result. Relations diagrams are used when the causes are non-hierarchical and when there are multiple interrelated problems; they tend to be used when there is a strong feeling that the problem under discussion is only a symptom. They allow the problem to be analysed from a wide perspective, as a specific framework is not employed, and allow for the use of multi-directional rather than linear thinking. Figure 7.12 examines the relationships between the issues that a manufacturing operation is experiencing.

Affinity Diagrams

This is used to categorise verbal and language data about previously unexplored issues, problems and themes which are hazy, uncertain, large, complex and difficult to understand, thereby helping to create order out of chaos. It is useful to use this after a brainstorming activity to order and categorise the ideas and suggestions that have been generated. This diagram uses the natural affinity between opinions and partial data from a variety of situations to help understand and structure the problem. It tends to be a creative rather than a logical process.

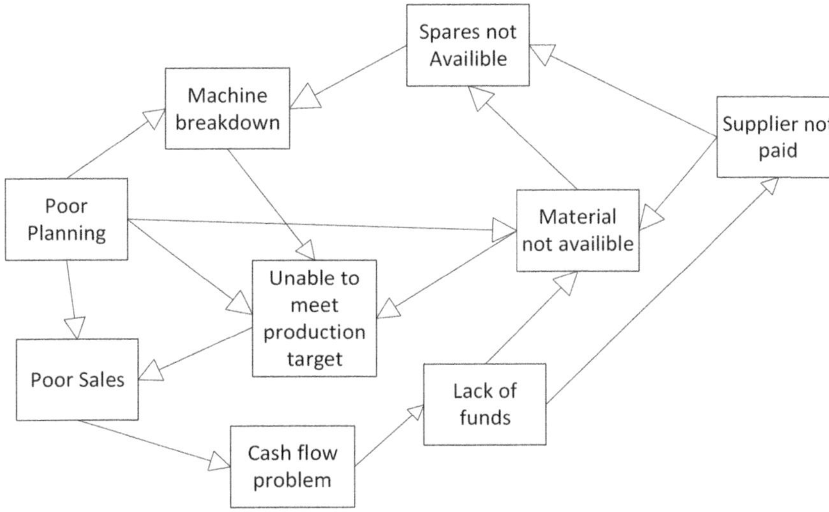

Figure 7.12 Example relationship diagram.

Figure 7.13 shows how the ideas generated during a brainstorming session exploring the issues affecting a food processing operation could be grouped.

Systematic Diagrams

Systematic diagrams, also called tree diagrams, are used to examine, in a systematic manner, the most appropriate and effective means of planning how to accomplish a task or solve a problem; events are represented in the form of a root-and-branch relationship. It displays in increasing detail the means and paths necessary to achieve a specific goal or to clarify the component parts which lead to the root cause of a problem. They are used when the causes that influence the problem are known, but a plan and a method for resolving the problem have not been developed. They can also be useful when a task has been simple but has run into implementation difficulties. A systematic diagram is usually used to evaluate several different methods and plans for solving a problem and thereby assist with complex implementation. It is used to identify dependencies in each situation and to search for the most suitable improvement opportunities, and also when there are major consequences for missing key tasks. Some of the issues that would need to be considered when exploring how to increase sales are illustrated in Figure 7.14.

Matrix Diagrams

This is used to clarify the relationship and connecting points between results and causes or between objectives and methods, and, using codes, to indicate their relative importance and the direction of the influence. The diagram is also useful to draw conclusions between consequences and their causes. They are used when there are two sets of factors and methods which may have no relationship with each other, and when there is a need to get a cumulative numerical score to compare one item to another. The factors are arranged in rows and columns on a chart with the intersections identifying the problem and its concentration; the

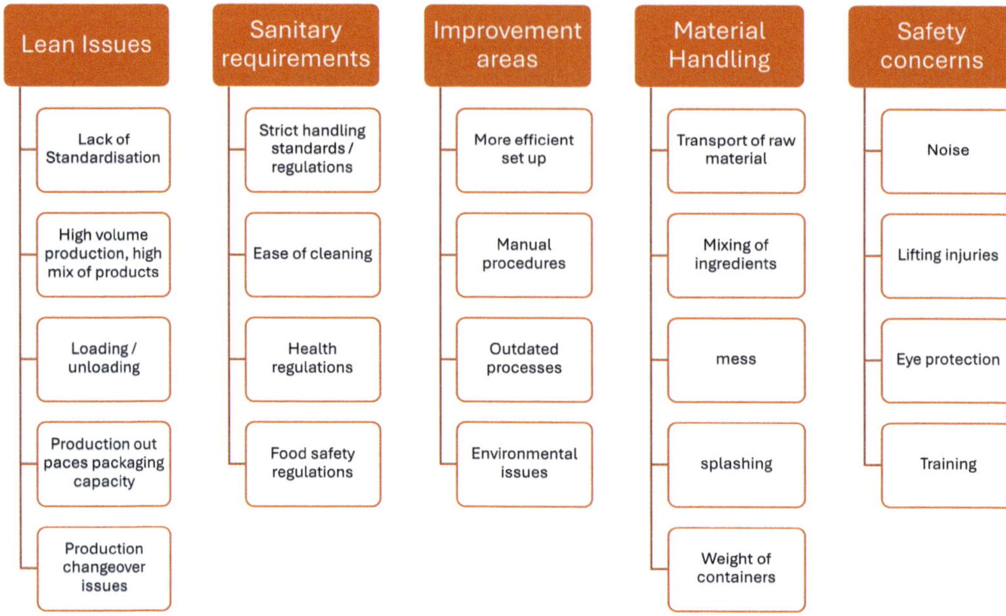

Figure 7.13 Affinity diagram for a food processing operation.

Figure 7.14 Example of a systematic diagram.

intersecting points are the basis for future action and problem-solving. Quality function deployment employs Matric diagrams to good effect.

Figure 7.15 displays some of the service features that a call centre may offer and their relative importance to different prospective customers. Through this type of analysis, the call centre can plan which of the customers to target their operation at and which features to offer.

Process Decision Programme Charts

Process decision programme charts (PDPC) are used to select the best processes to obtain the desired outcome from a problem statement by evaluating all possible events, contingencies and outcomes that can occur in any implementation plan. Considering the system, it is used to anticipate unexpected events and develop plans, countermeasures and actions for such outcomes. It is used to plan each possible chain of events that might occur when the problem or goal is unfamiliar, new, or unique, particularly when the stakes of potential failure are high. In this it is like FMEA and fault tree analysis. However, it is claimed to be more dynamic than these two methods since the relationship between the initiating condition/event and terminating condition/event has been thought out and mapped. It is based on a systematic diagram and uses a questioning technique – for example, 'What could go wrong?', 'What are the alternatives?' – and lists actions or countermeasures accordingly. The PDPC has no prescribed set of rules. A partial PCDC analysing how to minimise waste of material when cutting it to a pattern is shown in Figure 7.16.

Arrow Diagrams

This method applies systematic thinking to the planning and execution of a complex set of tasks. It is used to establish the most suitable plan and schedule for a series of activities in a project, and to monitor its progress in an efficient manner to ensure adherence to the schedule. Arrow diagrams are necessary to describe the interrelationship and dependencies of tasks within a complex job or project. They are deployed at the implementation planning stage of a project. The sequence of the steps involved and their relation to each other are indicated by an arrow, and in this way a network of activities is developed. This method, its form of construction, calculations and identification of the critical path are well known and used in project management in relation to critical path analysis (CPA) and PERT. Figure 7.17 shows an example of an arrow diagram, a CPA chart for the implementation of a new IT system.

		Importance		
		Customer 1	Customer 2	Customer 3
Service feature	Low waiting time	9	7	8
	Call resolved quickly	7	6	9
	No need to transfer call	4	6	5
	Cost of service	9	9	7

Figure 7.15 Example matrix diagram for a call centre service.

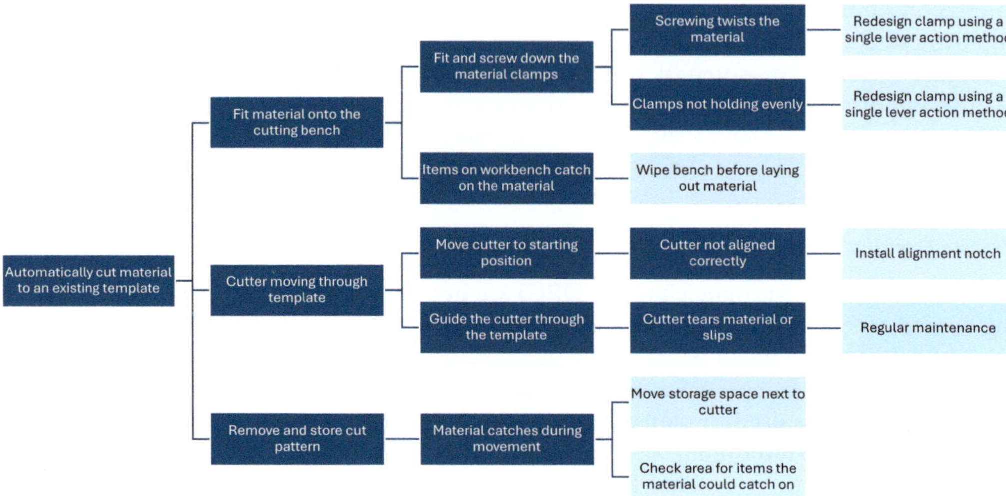

Figure 7.16 Process decision programme chart.

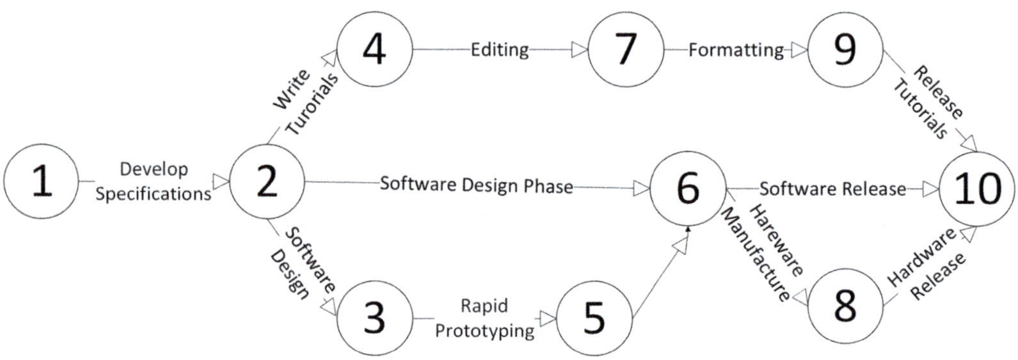

Figure 7.17 Arrow diagram (Critical path analysis for IT implementation).

5S Technique

The discipline of cleanliness and housekeeping is a prerequisite for effective quality assurance. It creates a working environment which is comfortable and in harmony with all employees. Organisations need to make sufficient effort to define and quantify their housekeeping requirements. It is also necessary to ensure that employees take responsibility for the cleanliness of the area where they work. A variety of tools can be used to promote housekeeping; one of the most common is the 5s approach, which organisations often implement as one of their first steps on their Lean journey. 5s is based around the Japanese concepts of seiri, seiton, seiso, seiketsu and shitsuke, which translate approximately as shown below.

- *Seiri*. Organisation: separating what is required from that which is not, eliminating those which are not required and/or tidying them away.

- *Seiton*. Neatness: arranging the required items in a tidy manner and in a clearly defined place, so that they can be accessed quickly.
- *Seiso*. Cleaning: keeping the surrounding area and environment clean and tidy.
- *Seiketsu*. Standardisation: clean the machinery and equipment according to laid-down standards and routines in order to identify deterioration.
- *Shitsuke*. Discipline: follow the procedures and the previous four steps which have been laid down and continuously improve them.

They sometimes also translate in English as: sort, set in order, shine, standardise and sustain (Minshull, Dehe, & Kotcharin, 2022).

Organisations have adopted this approach in different ways. For example, an NHS Hospital Operating Theatre Suite applied this to its main stock room in the following way:

- Sort – remove out-of-date and no longer used equipment, removing any items that should not be in the storeroom.
- Set in order – identify the commonly used items, improve shelving, store things by type and frequency of use.
- Shine – increase the cleaning, maintenance, and upkeep of the storeroom.
- Standardise – clearly label stock and quantities on the shelves.
- Sustain – include upkeep of the room.

Mistake-Proofing

Mistake-proofing is a technique which is used to prevent errors from being converted into defects. The concept was developed by Shingo (1986). The technique assumes that, no matter how observant or skilled people are, mistakes will occur unless preventative measures are put in place. Shingo argues that instead of looking for and correcting the causes of defective work, the source of the mistake should be inspected, analysed and rectified. He places great emphasis on what he calls source inspection, which checks for factors which cause mistakes, and then on using poka yoke or mistake-proofing devices to prevent their recurrence.

Mistake-proofing has two main steps: preventing the occurrence of a defect and detecting the defect. The purpose is to stop processes from operating when conditions exist that will lead to defects. The system is applied at three points in the process:

- In the event of an error, prevent the start of a process and shut it down.
- Prevent a non-conforming product from leaving a process.
- Prevent a non-conforming part from being passed to the next process.

The mistake-proofing technique employs the ingenuity and skills not only of the engineers and/or technical specialists who may develop and fit the devices, but also of the operators who have first identified the cause of the mistake and participated in the corrective action measures. Mistake-proofing devices, which may be as simple as a mechanical counter which ensures that the correct number of parts are fed into a machine, or cut-off switches, limit switches or float switches which provide some regulatory control of the process or operation, thereby stopping a machine or process automatically, should be used wherever possible. There may also be devices which prevent a part from being incorrectly fed into the machine, assembled incorrectly, fabricated incorrectly, or placed incorrectly into a fixture.

In other words, the assumption is made that, if the part can possibly be fed in wrongly, it will be unless some preventative measure is taken. This is the essence of mistake-proofing.

It is usual to supply the mistake-proofing device and signal with some audible or visual display or mimic diagrams and/or warning light to indicate that something has gone wrong, and not to plan to bring the abnormality to the attention of the operator. They are a relatively cheap and effective way of preventing defects.

Organisations that are looking to develop their own approach to mistake-proofing should consider the following:

- Mistake-proof at the earliest possible opportunity, ideally at the development stage and before any pre-production activities are undertaken.
- Involve manufacturing and quality department personnel in the research and development activity and ensure that there is a forum for the discussion of manufacturing and design problems and their interfaces; cross-functional teams, and concurrent and simultaneous engineering should facilitate this.
- The design and process FMEA, analysis of customer reject returns, warranty claims, field failure reports, in-house scrap and rework, and inspection data should help to pinpoint potential problems that could be resolved by mistake-proofing.
- It is much easier to mistake-proof new products than to develop devices for existing products.
- A team approach should be taken to study potential problems and likely causes of mistakes, and the development of mistake-proofing ideas and devices. The team should be multi-disciplinary and involve operators. Customers should also be involved, as this helps to build relationships and provides concrete proof that long-lasting improvement action is being taken. However, some suppliers are sensitive to their problems being exposed to customers.
- There should be some basic training in the principles, techniques, applications, and use of mistake-proofing, as well as other activities such as problem-solving and team building.
- To broaden the experience of mistake-proofing techniques and applications, information should be shared with other companies using the concept.

Total Productive Maintenance

The Japanese developed the concept of total productive maintenance (TPM), based on the planned approach to preventative maintenance (PM). TPM combines PM with QM and employee involvement and is considered a total method of management.

TPM is a scientific, company-wide approach in which every employee is concerned about the maintenance, quality and efficiency of their equipment. The objective is to reduce the whole-life cost of machinery and equipment through more efficient maintenance management, and as far as possible to integrate the maintenance and manufacturing departments. Teamwork is a key element of TPM. By analysing each piece of equipment, it focuses on reducing manufacturing losses and costs (i.e. the six major losses: breakdown, set up/adjustment, speed, idling and minor stoppages, quality defects and start-up; see Nakajima 1988), and establishes a system of PM over a machine's working life. The emphasis of TPM is to improve the skills of operators in relation to machine technology and to train and educate them to clean, maintain and adjust their machines. The training and education of operators is carried out by maintenance and engineering staff. In this way machinery is kept at optimal operating efficiency. The 5s are essential activities in TPM, and they also promote visible management.

The main organisational characteristics of TPM are:

- Integration of maintenance and production departments
- Small teams of operators/maintainers
- Training is undertaken to make operators feel like owners
- Good habits are developed:
  - Cleaning becomes checking
  - Cleaning highlights abnormalities
  - Abnormalities are rectified
  - Continuous improvement of the environment and equipment

The seven key steps of a TPM programme are usually considered to be:

Step 1 Initial cleaning is performed to identify problems with equipment that are not noticed during normal operations.
Step 2 Countermeasures at the source of problems, to minimise the accumulation of dirt and other contaminants and put in place improvements to make it easy to access parts of the equipment which need cleaning.
Step 3 Set maintenance, cleaning and lubrication standards for groups of equipment and carry out appropriate training.
Step 4 General inspection procedures and schedules.
Step 5 Autonomous inspection procedures and schedules.
Step 6 Orderliness and tidiness.
Step 7 Full autonomous maintenance.

Summary

Irrespective of the QM or strategic improvement approach chosen and followed, an organisation will need to use a selection of tools and techniques to assist with the process of continuous improvement. It is recommended that the simpler tools and techniques, such as the seven quality control tools, be used in the beginning, and that it be ensured that the tools and techniques which are currently employed are used effectively before attempts are made to introduce other techniques. A planned approach for the application of tools and techniques is necessary.

The temptation to single out one tool or technique for special attention should be resisted, and to get maximum benefit from the use of tools and techniques, they should be used in combination. It should be recognised that tools and techniques play different roles, and management and staff should be fully aware of the main purpose and use of the tools and techniques they are considering applying in the organisation; if this is not the case, they could well be disappointed if a tool or technique fails to live up to expectations. It is also important to understand the limitations of how and when tools and techniques can best be used. The tools and techniques should be used to facilitate improvement and be integrated into the way the business works rather than being used and viewed as 'bolt-on' techniques. The ways in which the tool or technique is applied and how its results are interpreted are critical to its successful use; a tool or technique is only as good as the person who is using it.

From research by Bamford and Greatbanks (2005) the following are key for the successful implementation, use and success of applying the QC and M7 tools and techniques:

1 In-depth knowledge of the process
2 Formal training in problem-solving techniques
3 Appropriateness of tools selected for use
4 Application simple models at all levels in the organisation to aid communication and learning

Tools and techniques on their own are not enough; they need an environment and technology conducive to improvement and to their use. An organisation's CEO and senior managers have a key role to play in the effective use of tools and techniques. They should, for example:

- Develop their knowledge of the tools and, when appropriate, use them in their day-to-day activities and decision-making.
- Be fully aware of the main purpose and use of the tools and techniques which are being applied.
- Delegate responsibility for their promotion to suitable individuals.
- Maintain an active interest in the use of tools and the results.
- Endorse expenditure arising from the education and training required and the improvement activities resulting from the employment of tools.
- Recognise and reward those employees who utilise tools and techniques in their day-to-day work activities.

References

Bamford, D., Forrester, P., & Reid, I. (2023). *Essential guide to operations management: Concepts & case notes*, 2nd edition. UK: Routledge.

Bamford, D., & Greatbanks, R. (2005). The use of quality management tools and techniques: A study of application in everyday situations. *International Journal of Quality and Reliability Management*, 22(4), 376–392.

Barker, R.L. (1989). The seven new QC tools. In *Proceedings of the first conference on tools and techniques for QM*, 95–120. Bedford: IFS Conferences.

Ishikawa, K. (1976). *Guide to quality control*. Tokyo: Asian Productivity Organisation.

Juran, J.M. (1988). *Quality control handbook*. New York: McGraw-Hill.

Mathur, S., Antony, J., Olivia, M., Fabiane Letícia, L., Shreeranga, B., Raja, J., & Ayon, C. (2023). An empirical study into the use of 7 quality control tools in higher education institutions (HEIs). *The TQM Journal*, 35(7), 1777–1798.

McDermott, O., Antony, J., Sony, M., Fernandes, M.M., Koul, R., & Doulatabadi, M. (2023). The use and application of the 7 new quality control tools in the manufacturing sector: A global study. *The QM Journal*, 35(8), 2621–2639.

Minshull, L.K., Dehe, B., & Kotcharin, S. (2022). Exploring the impact of a sequential lean implementation within a micro-firm: A socio-technical perspective. *Journal of Business Research*, 151, 156–169.

Mizuno, S. (1988). *Management for quality improvement: The seven new QC tools*. Cambridge, Mass.: Productivity Press.

Nakajima, S. (1988). *Introduction to total productive maintenance*. Cambridge, Mass.: Productivity Press.

Ozeki, K., & Asaka, T. (1990). *Handbook of quality tools*. Cambridge, Mass.: Pro- ductivity Press.

Shingo, S. (1986). *Zero quality control: Source inspection and the Poka Yoke System*. Cambridge, Mass.: Productivity Press.

8 Quality Management Techniques

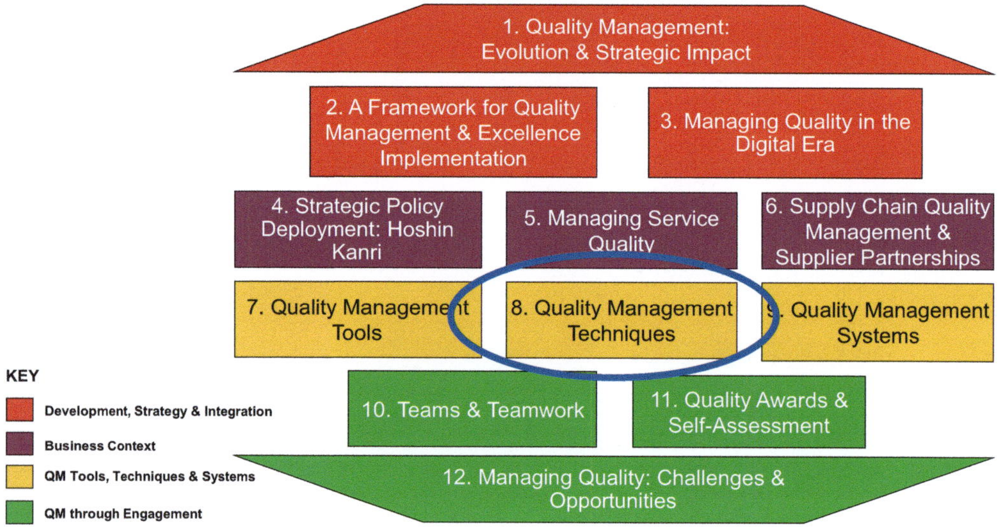

Introduction

This chapter provides an overview of the core quality management techniques and of 'six-sigma', a strategic improvement approach, often deployed in an organisation's improvement process.

1 Quality Function Deployment
2 Design of Experiments
3 Failure Mode and Effects Analysis
4 Benchmarking
5 Value Stream Mapping
6 Six-Sigma

Previous editions of this textbook included a detailed discussion of Business Process Re-engineering. This, as a concept, is now becoming rather dated, and although it has evolved over the years, several tools and methodologies have started to supersede it. Some of the notable ones include:

DOI: 10.4324/9781003594369-11

- Lean Six-Sigma: This combines Lean manufacturing principles with six-sigma methodologies to reduce waste and improve process quality.
- Agile Methodology: Originally developed for software development, Agile principles are now applied across various industries to enhance flexibility and responsiveness.
- Business Process Management (BPM): BPM involves the use of various tools and techniques to model, analyse, and optimise business processes.
- Kaizen: A Japanese term meaning 'continuous improvement', Kaizen encourages small, incremental changes to improve efficiency and quality.

These tools and methodologies help organisations streamline operations, reduce costs, and enhance overall quality.

Quality Function Deployment

Quality Function Deployment (QFD) is a systematic procedure which is used to help build quality into the upstream processes and new product development. It helps to avoid problems in the downstream production and delivery processes and will consequently shorten the new product/service development time. The concept helps to promote proactive rather than reactive development by capturing and measuring the 'voice of the customer'.

QFD is a systematic technique which is used to translate the needs of the customers into design requirements, technical requirements, and operational terms (Bamford, Forrester, & Reid, 2023; Hariri, Domingues, & Sampaio, 2023). It is based on the philosophy that the 'voice of the customer' drives all company operations and requires reliable data from the following diverse sources: Customers; Design functionality; Costs and capital; Reliability; Reproducibility.

QFD seeks to identify those features of a product or service which satisfy the real needs and requirements of customers (market- or customer-required quality), and can be used systematically to guide the company on how best to enhance performance and customer satisfaction (Kassela, Papalexi, & Bamford, 2017). A critical part of the analysis is that it takes into account discussions with the people who actually use the product in order to obtain data on issues such as:

- What do they feel about existing products?
- What bothers them?
- What features should new products have?
- What is required to satisfy their needs, expectations, thinking, and ideas?
- How and where is the product used?

Understanding Customer Needs

The voice of the customer is the cornerstone of QFD. Hence, talking and listening to the customer is paramount to understanding their real needs and requirements. The preferred method of doing this is through direct contact with the customer, utilising customer questionnaires, face-to-face discussions with customers, or other forms of direct consumer contact. Other methods that might be used to understand customer needs include: Failure-related information, which shows where customer needs are not being met, for example, field-failure data, warranty returns, or customer complaints; and Data from various sources, for example, market surveys, dealer information, trade shows, test marketing, etc.

In using these methods for understanding customer requirements, typical issues that need to be considered include:

- What is wrong with the product and/or service
- Performance features that delight the customer
- The 'if only' factor
- When the product is used
- How the product is used
- Who uses the product

As can be seen from the above discussion, QFD is useful in terms of:

- Building the 'voice of the customer' into the design process
- New product development projects
- Process design, redesign, and improvement projects

How to Use Quality Function Deployment

The main objectives when using this technique are to:

- Identify customer requirements
- Determine competitive opportunities
- Determine substitute quality characteristics
- Pinpoint requirements for further study.

An example of the House of Quality derived from the product-planning phase of QFD is shown in Figure 8.1. In this example, the key elements of the process comprise the following:

- WHATs: the wants and needs of the customer – the Voice of the Customer.
- HOWs: The design features and the technical requirements to deliver the WHATs
- WHATs vs HOWs: the HOWs are ranked according to their effectiveness in delivering each of the WHATs. Usually, a ranking system is used to indicate a strong, moderate or weak correlation.
- HOWs vs HOWs: This is usually displayed as a matrix which visualises how the design features interact with each other. Again, a ranking system is used, ranging from strong positive to strong negative interrelationships.
- HOWs vs HOW MUCHes: these are the target or benchmark values for each of the HOWs
- WHATs vs WHYs: these are the justifications for each of the WHATs

The Project

The scope of the project should be clearly outlined, including targets, operating constraints and timescale. A clearly defined mission statement should be produced, and a team formed. It is useful to create a business model which includes market definition and size, product life

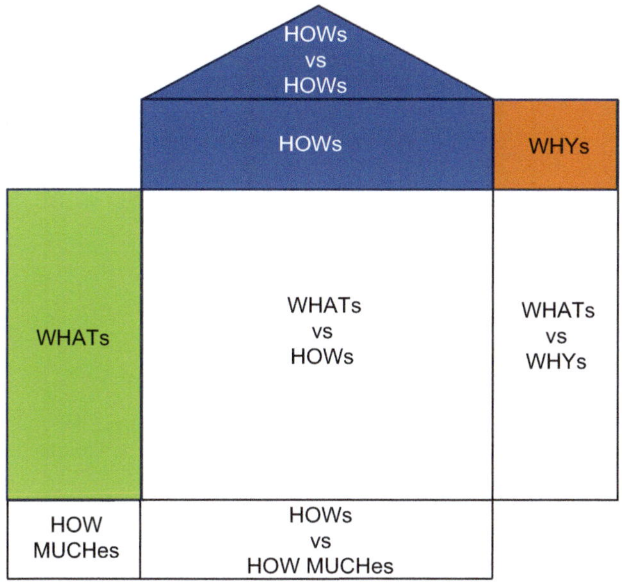

Figure 8.1 The house of quality (Bamford et al., 2023).

history, competitive products and prices, projected sales, prices and costs, and the estimated capital requirements and likely payback.

Customer Needs

Gathering the voices of customers can be done in different ways, as previously detailed. The information gathered can be entered into a chart similar to that shown in Figure 8.2, complete with full information on why the product is needed, for what purposes, who uses it and when, and where and how it is used. This information provides the basis for more easily translating the customer's voice into customer needs, which can be satisfied by design features. For example, 'In the UK, people commonly use mobile phones while on the move' will translate the needs of customers into a requirement for one-handed operation of the phone, including the ability to dial and hold from the same hand (WHATs). The phone will then conveniently have design features of the width and depth of the mobile phone, button areas and depression forces, etc. (HOWs), See Figure 8.1.

Implementation of Quality Function Deployment

The following are key steps in the effective implementation of QFD:

- *Management issues*
  - The process must be driven by senior management.
  - Appropriate resources and the provision of training need to be allocated and actioned by management.
  - Appoint a steering committee and a QFD champion.
  - Use of a project management system to act as a communication vehicle.

Customer Classification:	Voice of the customer: Actual information (the spoken or written words)					
Internal / External **Socio Economic Group:** **Age:** **Sex:** 1 2 3 4 5 6 7	**WHY** do you want or need this product?	**WHAT** is it or will it be used for?	**WHO** uses or will use it?	**WHEN** do or will customers use it?	**WHERE** do or will customers use it?	**HOW** is or will the product / service be used?

Figure 8.2 Gathering the voice of the customer and interpreting it into customer needs.

- *Project issues*
  - Select the first project with a limited timeframe and a good chance of an early success.
  - Establish a timeframe for the project at the outset and keep to it.
  - Have a clear project definition and objectives, and always have them in view; it is also important to identify any project limitations and operating constraints. This helps to create a focus on what is being done.
  - Develop a clear market definition and business model.
  - Provide a glossary of terms used in the QFD process.
- *The QFD team*
  - Train as a team using as many company-specific examples as possible.
  - Establish a core team, which is multidisciplinary, of between five and seven people.
  - Hold short, regular team meetings.
  - Do detailed work outside the meeting and use the meeting for analysis and decision-making. It is important that each member of the team is prepared to make a significant time commitment to the project.
  - One of the cornerstones of QFD is the customer's voice – it is best for all the team to be part of the data-collection process, which involves listening to that voice.
  - It takes longer, but consensus decisions generally work best.
  - Team energy can be created by paying attention to: direction, structure, project management, and human issues.
- *Methods of working*

- Do as much concurrent work as possible (e.g. competitive benchmarking with existing product and process designs).
- Create a planning matrix of customer needs to decide what should go into a house of quality; some items will be better achieved by traditional means.
- Keep a realistic perspective on the details entered about customer needs. Focus on the important, the difficult, and the new.
- Try to ensure that the house of quality is kept to within an approximately 30 × 30 matrix.
- Use the customer's voice and benchmarks as major decision-makers to achieve best in class.

Advantages of Quality Function Deployment

QFD offers an organisation a number of benefits (Dehe and Bamford, 2017):

- It converts the 'voice of the customer' into design requirements which can then be utilised in new product development and process design projects.
- It benchmarks the organisation's products and services against competitors.
- Reduces the cost of new products.
- Speeds up the design process.
- Increases customer satisfaction.

Disadvantages of Quality Function Deployment

There are a number of issues that have been identified with QFD (Dehe and Bamford, 2017):

- A large amount of data is required.
- It can take a large amount of time.
- It assumes that customer requirements do not change over time.

Summary of QFD

The QFD process provides a powerful structure for product and process development. When it is used in an effective manner, it can bring a correct customer focus to designs that will perform to a high degree of satisfaction with reliability and cost-worthiness. It shortens the development cycle for the design and results in fewer engineering changes. In this way the product/service which the customer receives not only meets their needs but, if the customer interface has been done correctly, there can be unexpected product features which will cause delight and product loyalty. There will also be a common thread through all operations, which is traceable back to what the customer really wants.

Design of Experiments

The Design of Experiments (DoE) is a series of techniques, which involve the identification and control of parameters which have a potential impact on the performance and reliability of a product design and/or the output of a process (Arboretti et al., 2022), with the objective of optimising product design, process design and process operation, and limiting the influence of 'noise' factors. The methodology is used to analyse the significance of effects on system outputs of different values of design parameters. The objective is to optimise the values of these design parameters to reduce or remove the variation from the performance of the system. There are several methodologies of experiments such as: The trial and error, the full factorial, the fractional factorial and the Taguchi method. In this section only an overview of the Taguchi technique is provided.

DoE can be applied to the design of new products and processes or to the redesign of existing ones, and is used to:

- Optimise product design, process design, and process operation.
- Achieve minimum variation of best system performance.
- Achieve reproducibility of best system performance in manufacture and use.
- Improve the productivity of design activities.
- Evaluate the statistical significance of the effect of any controlling factor on the outputs.
- Reduce costs.

How to Use Design of Experiments

There are several methodologies for DoE, such as: The trial and error, the full factorial, the fractional factorial and the Taguchi (1986) method. In this section we will focus on the Taguchi technique.

DoE historically required a great deal of statistical knowledge and understanding, which most industrial users of experiments found somewhat intimidating. Over the years much effort has been devoted to simplifying the task of experimentation. In the late 1970s, the work of Genichi Taguchi on experimental design made what is regarded by many as a major breakthrough in its application. Dr. Taguchi was a statistician and electrical engineer who was involved in rebuilding the Japanese telephone system and was involved in applying the DoE in the Japanese electronics industry for over 30 years. He promotes three distinct stages of designing in quality:

1 *System design*: the basic configuration of the system is developed. This involves the selection of parts and materials and the use of feasibility studies and prototyping. In system design technical knowledge and scientific skills are paramount.
2 *Parameter design*: the numerical values for the system variables (product and process parameters – termed factors) are chosen so that the system performs well, no matter what disturbances or 'noises' (i.e. uncontrollable variables) are encountered by the system (i.e. robustness). The objective is to identify optimum levels for these control factors so that the product and/or process is least sensitive to the effect of changes in noise factors. The experimentation pinpoints the best combination of product/process parameter levels. The emphasis in parameter design is on using low-cost materials and processes in the production of the system; it is a key stage of designing for quality.
3 *Tolerance design*: the third stage in the design process uses experimental design to investigate the effect on the variance of the output characteristic of:

   – Product design: choosing the upper specification limit (USL) and lower specification limit (LSL) around the nominals of key design parameters that have been prescribed by the parameter design study. Having done this, reconciling the choice of limits of

the factors in the design that are predicted to cause most variation, with, typically, the cost of reducing the tolerance gap, or the choice of more expensive materials.
- Process design: choosing the USL and LSL around the nominals of key process factors that have been prescribed by the parameter design study. Having done this, reconciling the choice of limits of the factors in the process that are predicted to cause most variation, with, typically, the cost of reducing the tolerance gap, or the choice of more expensive methods.

Taguchi's approach also addresses the following:

- Determining the quality level, as expressed in the loss function concept.
- Improving the quality level in a cost-effective manner by parameter and tolerance design.
- Monitoring the quality level using Statistical Process Control (See later in the chapter). A feedback/feed-forward closed-loop system is also recommended.

Taguchi's methods (i.e. engineering, experimental design, and data analysis) have proven successful both in Japan and the West, and those organisations which have adopted his methods have succeeded in making continuous improvement. There is little doubt that his work has led to increased interest by engineers in a variety of approaches and methodologies relating to the DoE. He has provided a technique to analyse the effects of control factors on variability with respect to noise. However, it should not be overlooked that a number of other people have made significant improvements with the other approaches to experimental design.

Steps in Experimental Design

The key steps in designing and running a DoE project are:

1 Define the project objectives
2 Select critical characteristics
3 Determine the issues that affect the critical characteristics
4 Identify control factors and 'noise' factors
5 Select the control factors to be optimised during the experiment
6 Choose the orthogonal array and assign factors to columns in the array
7 Choose the levels of the control factors
8 Choose sample size
9 Organise the experiment and carry it out
10 Analyse the data
11 Predict the result of the confirmation run
12 Interpret the confirmation run and decide if the project is finished

Advantages of Design of Experiments

DoE offers a number of benefits for product and service design or improvement:

- Experimental design makes it possible to determine a cause-and-effect relationship between the control factors and the results of the experiment.

- It facilitates the elimination of 'noise' factors (unwanted extraneous variables).
- Due to rigorous experiment design, experiments can be repeated to verify the results.

Disadvantages of Design or Experiments

There are a number of potential drawbacks to DoE:

- It requires training in statistical and scientific techniques.
- It requires a great deal of technical knowledge about the products, systems and technologies that are being experimented on.

Summary of DoE

Experimental design using a variety of matrices which suit different conditions is a key technique for understanding the effect of each controllable factor, be it a product or a process design, in minimising variation while focusing the output on a target value. It is a major technique in investigating quality problems. Statistical DoE is a complex subject, but it is possible to develop 'easy-to-use' methods. .

Failure Mode and Effects Analysis

This section provides an overview of the concept of failure mode and effects analysis (FMEA), and its value as a planning tool to assist with building quality into an organisation's product, service and processes.

The technique of FMEA was developed in the 1960s in the aerospace and defence industries as a method of reliability analysis, risk analysis and risk management. It is a systematic and analytical quality planning tool for identifying, at the product, service and process design and development stages, what might potentially go wrong, either with a product (during manufacture, or during end-use by the customer), or with the provision of a service, thereby aiding fault diagnosis. The use of FMEA is a powerful aid to advanced quality planning of new products and services, and can be applied to a wide range of problems which may occur in any system or process.

The usefulness of FMEA as an approach is reflected in its adaptation and adoption as an effective framework for risk management and mitigation in project management.

As discussed above, FMEA can be used to:

- Identify, manage and mitigate against risks in product design, system design, and project management.
- Facilitating the identification, diagnosis, and fixing of faults.
- Build quality into products and systems.

How to Use Failure Mode and Effects Analysis

There are two categories of FMEA: design and process. A design FMEA assesses what could, if not corrected, go wrong with the product in use and during manufacture because of a weakness in the design. Design FMEA also assists in the identification or confirmation of critical characteristics. On the other hand, process FMEA is mainly concerned with the reasons for potential failure during manufacture and in use as a result of non-compliance with the original design intent, or failure to achieve the design specifications.

FMEA explores ways in which a product, service or process can fail and is best carried out as an iterative process. In brief, it involves the following steps:

- The function of the product, service and/or process is agreed, along with suitable identifications.
- Potential failure of the product or service is identified.
- The effects of each potential failure are assessed and summarised.
- The causes of the potential failures are examined.
- Current controls for the detection of failures are identified and reviewed.
- A Risk Priority Number (RPN) is determined; the details are provided below.
- The corrective action which is to be taken to help eliminate potential concerns is decided.
- The potential failures in descending order of RPN are the focus for improvement action to reduce/eliminate the risk of failure occurring.
- The recommendations, corrective actions, and countermeasures which have been put into place are monitored and reviewed for effectiveness.

The RPN comprises an assessment of occurrence, detection, and severity of the potential failure and is the multiplication of the three rankings:

1 The *occurrence* is the likelihood of a specific cause actually happening, which will result in the identified failure, and is based on perceived or (in the case of process capability) estimated probability of occurrence. It is ranked on a scale of 1–10.
2 The *detection* criterion relates, in the case of a design FMEA, to the likelihood of the design verification programme pinpointing a potential failure before it reaches the customer; a ranking of 1–10 is again used. In the process of FMEA, the detection criterion relates to the existing control plan.
3 The *severity of effect*, on a scale of 1–10, indicates the likelihood of the customer noticing any difference in the functionality of the product or service.

The resulting RPN should always be checked against past experience of similar products, services and situations.

The requisite information and actions are recorded in a standard format in the appropriate columns. An example of a process FMEA for automobile assembly is shown in Figure 8.3. The FMEA is a live document and should always be modified in light of new information or changes.

From the design FMEA, the potential causes of failure should be studied and actions taken before designs and drawings are finalised. When used in the proper manner, FMEA prevents potential failures occurring in the manufacturing, production and/or delivery processes or end-product in use, and will ensure that processes, products and services are more robust and reliable. It is a powerful technique, and several well-publicised product recall campaigns could conceivably be avoided by the effective use of FMEA. However, it is important that FMEA is seen not just as a catalogue of potential failures, but as a means for pursuing continuous improvement. Nor should it be viewed as a paperwork exercise carried out merely to retain business, as this will limit its usefulness. The concept, procedures and logic involved with FMEA are not new: every forward-thinking design, planning

Process Function Requirements	Potential Failure Mode	Potential Effect(s) of Failure	Sev	Class	Potential Cause(s)/ Mechanism(s) of Failure	Occur	Current Process Controls	Detec	R.P.N.	Recommended Action(s)	Responsibility & Target Completion Date	Action Results				
												Actions Taken	Sev	Occ	Det	R.P.N.
Kit parts to the Assembly line	Incorrect parts used on build	Turbo failure	7		Wrong parts presented to the line at change over	2	Visual check by setter to the shop Packet. Introduced from the bulk issue area set up on the line Marking of the part number on the 'A' surface on Compressors to ID. Marking of the part number on the Turbine flange.	3	42							
Kit parts to the Assembly line	Contaminated parts	Turbo failure	7		Contaminated parts due to lack of cleanliness of holding containers, Organic material in spacer (GT)	2	Visual check, work instructions by Station describing method of Assembly.	7	98	Introduce cleaning process for all the boxes, which the parts are presented to the line. Euroboxes 300mm × 200mm, 400mm × 300mm & 600mm × 250mm	LM, DN, PE, BW, mid December 20XX	Quotes collected, Capital approval sanctioned Delivery 12 December 1997	7	2	3	42
Check cross over holes	Cross over hole not drilled	no oil flow Turbo failure	8		Broken drill, missed operation	5	Air Gauge on assembly 100% prior to build; work instructions by Station describing method of Assembly.	3	120	New poka yoked fixture provided for an 'in process' end of line check. Will pressure test the Centre Housing and check that the cross over hole is present. If OK a letter 'T' will be stamped.	PC, DC, MB Oct 20XX	New end of line test fixture, pressure tests CH and checks for Cross over hole, being debugged at supplier to be on stream 1st Dec 1997	8	2	3	48
Affix label to Centre Housing	Wrongly orientated to Customer requirements or on Wrong side of Centre Housing	Unable to read the customer no on the engine, reject unit which customer will have to adjust label to correct orientation and record ppm	5		Process controlled by the operator and is capable of producing defects	4	100% Visual check by operators; work instructions by Station describing method of Assembly.	5	100	Design new fixture to mistake proof, by Interlocking the fixture to prevent Assembly. Will prevent the Stick Screws from being supplied with air on the detection of incorrect label orientation.	LM, AM, FW, 15 November 20XX	Design and detail drawings being modified to suit CH with Backplate assembled	5	2	3	30
Affix label to Centre Housing or Compressor Hsg	Incorrect data on label	Customer Dissatisfaction Could use the wrong turbo	7		Wrong input	2	Software provides for a checksum so that the data has to be inputted twice to verify; work instructions by Station describing method of Assembly.	5	70							
Affix label to Centre Housing or Compressor Hsg	Label not properly affixed	label will fall off	5		Hole oversize from machining, stripped thread	4	100% Visual check by operators when recording the Serial Number on audit sheet; work instructions by Station describing method of Assembly.	3	60							
Affix label to Centre Housing or Compressor Hsg	Label Missing	Unable to ID unit	5		Operation carried out incorrectly	4	100% Visual check by operators Serial Number recorded on audit sheet; work instructions by Station describing method of Assembly.	2	40							

Figure 8.3 Potential failure mode and effects analysis (process FMEA).

(Adapted from Dale et al., 2016.)

and production engineer, technical specialist, and project manager carries out, in an informal manner, various aspects of FMEA. In fact, most of us in our daily routines will subconsciously use a simple informal FMEA. However, this mental analysis is often not committed to paper in a format which can be evaluated by others and discussed as the basis for a corrective action plan. What FMEA does is to provide a planned, systematic method of capturing and documenting this knowledge. It also forces people to use a disciplined approach and is a vehicle for obtaining collective knowledge and experience through a team activity.

Development of a Design FMEA

For a design FMEA the potential failure may be caused, for example, by an incorrect material choice, part geometry, or inappropriate dimensional specification.

The procedure then identifies the effects of each potential failure mode, examines the causes of potential failure and reviews current controls for the design FMEA, which usually include some form of design verification programme. For example, in the case of a turbocharger this includes items such as material pull tests, heat-cycling tests of components subject to high temperatures, life-cycle fatigue tests to failure, static engine testing, and dynamic engine testing on development vehicles. Regarding the latter, these tests are often carried out by the customers as part of their overall engine/vehicle evaluation programme. Experience with similar products is often used to verify the validity of certain component parts for a design.

The occurrence for a design FMEA is an estimate, on a scale of 1–10, of the potential failure occurring at the hands of the customer, a ranking of 1 indicating that the failure is unlikely (typifying a possible failure rate of <1 in a million), and a ranking of 10 indicating an almost inevitable failure (typically 1 in 2).

The detection criterion rests on the likelihood of a current design verification programme highlighting a potential failure mode before it reaches the customer. A ranking of 1 indicates almost certain detection, and a ranking of 10 indicates that the current controls are very unlikely to detect the failure mode before dispatch to the customer.

The severity-of-effect ranking is again on a 1–10 basis. A ranking of 1 indicates that the customer is unlikely to notice any real effect, in the case of a vehicle, on performance or the performance of the sub-system. A ranking of 10 implies that a potential failure could affect safe vehicle operation and/or non-compliance with government regulations. A severity ranking cannot be altered by any means other than by redesign of a component or assembly; it is a fixed feature. Clearly, serious implications exist under product liability legislation for high-severity rankings, and these high rankings must be addressed as a matter of urgency.

Development of a Process FMEA

In the case of a process FMEA the potential failure mode may be caused by, for example, the operator assembling the part incorrectly, or by variation in the performance of the equipment or data entered incorrectly into a system by an operator.

The procedure then, as in the case of a design FMEA, identifies the effects of each potential failure mode, examines the causes of the potential failure mode, and reviews the current controls. For a process FMEA the current controls might be operator-performed inspection or Statistical Process Control (discussed in the next section of the chapter), providing information on the capability of the process. The occurrence of a process FMEA is again based on a 1–10 scale, with a ranking of 1 indicating that a failure in the manufacturing process is almost certain not to occur. This is based on the experience of a similar process, both within the factory and in the field with the customer, typically identified by a high process capability value. Conversely, a ranking of 10 indicates that the failure is almost certain to occur and will almost definitely reach the subsequent operation or customer if countermeasures and controls are not put in place. An occurrence ranking of 10 suggests, and indeed demands, that corrective action be undertaken because it highlights a potentially incapable process.

Detection rankings for a process FMEA indicate a ranking of 1, indicating that the potential failure is unlikely to go undetected through the manufacturing process or service delivery. A ranking of 10 suggests that current monitoring and measurement procedures are unlikely to detect the potential failure in the component or the assembly before it reaches the customer, and that urgent corrective action is required. It is interesting to note that a successive inspection check (e.g. bolt torque conformance) does not result in the detection ranking being markedly reduced; it would still be assigned a ranking of between 7 and 10 since experience indicates that 100 per cent subsequent inspection is only capable of detecting 80 per cent or so of defects. The situation would be assessed differently in the case of automated inspection.

A much better method of detection is to introduce a successive check at a subsequent operation, whereby the operator is unable to perform his or her operation unless the previous operation has been correctly executed. This can be achieved by designing fixturing in such a way that it will only accept conforming parts from a previous operation. Another method is to install error-proofing devices at the source.

The criterion for the severity-of-effect ranking is determined in a similar manner to that for a design FMEA. The activity following the evaluation of current controls for a process FMEA is again the determination of the RPN.

Analysis of Failure Data

To apply FMEA effectively it is necessary to obtain real-world figures for the calculation of the RPN, for internal and external failure rates. These can then be used for the compilation of the occurrence ranking. External failure data can be collated using computer aids by field service engineers, customer return data, and customer surveys. The data obtained from visits to customers to review products and services which have failed, the disposition of which is determined (i.e. whether the failure liability is due to the manufacturer / service delivery organisation or the customer or if, in some cases, there is in fact no fault found). Internal process failure rates are collated weekly by the quality assurance department. The data are obtained from rejection notes associated with non-conforming activities and parts by quality assurance personnel.

It is important to realise that if a process FMEA is being compiled for a new process, for which no internal or external failure rate history is known, then it is acceptable to use judgement on failure rates for a similar process. For example, external failure rates over a five-year period could be collated and then ranked highest to lowest to identify the highest occurring items. Consideration must also be taken of any high-occurrence failures attributable to one cause. An investigation should be undertaken to see if the cause has been eliminated and, if so, then these should be ignored. The emphasis should be placed on identifying consistent patterns of regularly occurring effects of failure. These types of failure are the ones to which corrective action should be applied.

Recommended Actions for Design and Process FMEA

Following the determination of the RPN it is usual to perform a Pareto analysis and address the potential failure modes in order of decreasing RPN. Determining the figure for an acceptable RPN is really a matter of the application of common sense. If 100 is assumed to be the acceptable maximum, then this should be checked against experience. The rule to be applied is to adopt a consistent approach for each of the rankings, and generally, it will be found that the high RPNs are as expected. This takes the form of identifying recommended action(s) and assigning personnel to take appropriate improvement measures by a particular date, which should be before the scheduled product release to the customer.

Following satisfactory completion of the actions the RPN can be recalculated, and if it is below the agreed acceptable limits, then the corrective action can be assumed to be adequate. If this is not the case, then the design or process must be readdressed by appropriate corrective actions.

For a design FMEA the potential failure causes must be studied before drawings are released to production status. In the case of the process FMEA, all the controls and measures to ensure design intent, which are carried forward into the final product or service, must be implemented. If this is not done properly, then problems relating to identified failure modes will occur during product or service use. In the case of a new process, potential failure modes may be overlooked because of a lack of experience. However, if this is discovered later, these must be included in both process and design FMEAs for future consideration.

Advantages of Failure Mode and Effects Analysis

Effective use of FMEA should lead to a reduction in:

- Defects during the production of initial samples and in volume production
- Customer complaints
- Failures in the field
- Performance-related deficiencies (these are less likely if a detailed development plan is generated from the design FMEA)
- Warranty claims
- Safety concerns

In addition, there will be improved customer satisfaction and confidence as products and services are produced from robust and reliable production and service delivery methods. It also has relevance in the case of product liability.

Disadvantages of Failure Mode and Effects Analysis

As you might expect FMEA is not without its limitations:

- FMEA can be difficult, laborious, and tedious to implement for the case of complex systems that have multiple functions involving different sets of system components.

- FMEA assumes that failures occur independently of each other. Relationships between individual failures or groups of failures or causes of failure cannot be effectively presented in FMEA.
- There is the tendency to rely on subjective rather than objective assessments to generate the RPN for individual failures.

Summary of FMEA

Finally, a few dos and don'ts are given, which may help organisations to avoid some of the difficulties and traps typically encountered in the preparation and use of FMEA.

Do

- Develop a strategy for the use of FMEA.
- Drive the implementation with the full support of senior managers; it is the responsibility of the senior management to see that there is a positive attitude in the organisation towards FMEA.
- Ensure that all personnel who are to be involved with the FMEA are made aware of the potential benefits arising from the procedure and the necessity for corrective action to be implemented if improvements are to be made.
- Try to ensure that specialist staff feel that FMEAs are an important part of their job.
- Make FMEA meetings short but regular throughout the early stages of the product life cycle.
- Consider producing FMEA for product/service families, material categories, main assemblies, services, and process routes (i.e. generic FMEA) rather than for each component.
- Put into place a procedure for review/update of the FMEA; it should always be treated as a living document.

Don't

- Overlook the benefits of involving customers and suppliers in the preparation of FMEA.
- Start the FMEA process when the design has reached an almost fixed state, when changes will be that much harder to affect.
- Allow the preparation of FMEA to be carried out in isolation by one individual.
- Allow important failure modes to be dismissed lightly with comments such as, 'we've always done it like this', or 'that will involve a considerable investment to change', without considering the feasibility and cost of the change.
- Use the technique as just window-dressing for the customer. There is little difference in the effort made when using FMEA in this way from that required when using it in the correct manner.

Benchmarking

From the late 1980s onwards there has been a growth of interest in the subject of benchmarking as part of the culture of continuous improvement. This has been triggered by the success of the improvement methods used by the Xerox Corporation and by the development of the self-assessment methods promoted by the MBNQA and EFQM models for business excellence. Benchmarking, as it is known today, originated in Rank Xerox. It is now well documented that when Rank Xerox started to evaluate its copying machines against the Japanese competition, it was found that the Japanese companies were selling their machines for what it cost Rank Xerox to make them. It was assumed that the Japanese-produced machines were of poor quality, but this proved not to be the case. This exposure of the corporation's vulnerability highlighted the need for change. In simple terms, the aim of benchmarking is to identify practices that can be implemented and adopted to improve company performance.

Benchmarking is an opportunity to learn from the experience of others. It helps to develop an improvement mindset amongst staff, facilitates an understanding of best practices and processes, helps to develop a better understanding of processes, challenges existing practices within the business, assists in setting goals based on fact and provides an educated viewpoint of what needs to be done rather than relying on whim and gut instinct.

Typically benchmarking is seen as a useful tool to identify performance gaps in relation to other high-performing companies which operate in the same industry (Twaissi et al., 2025). It is particularly useful for:

- Identifying opportunities for improvement
- Identifying the ways in which improvement might be achieved
- Better understanding your competition
- Improving product and service quality

How to Use Benchmarking

Most organisations carry out what can be termed informal benchmarking. This traditional form of benchmarking has been carried out for years, beginning with military leaders. It takes two main forms:

- Visits to other companies to obtain ideas on how to facilitate improvement in one's own organisation.
- The collection, in a variety of ways, of data about competitors.

This is often not done in any planned way; it is interesting but limited in its value owing to a lack of structure and clear objectives. This approach is often branded 'industrial tourism'. To make the most effective use of benchmarking and use it as a learning experience as part of a continuous process rather than a one-off exercise, a more formal approach is required.

There are three main types of formal benchmarking:

1 *Internal benchmarking*. This is the easiest and simplest form of benchmarking to conduct and involves benchmarking between businesses or functions within the same group of companies. Many companies commence benchmarking with this form of internal comparison. In this way the best internal practices and initiatives are shared across the corporate business.
2 *Competitive benchmarking*. This is a comparison with direct competitors, whether of products, services or processes within a company's market. It is often difficult, if not impossible in some industries, to obtain the data for this form of benchmarking, as by the very nature of being a competitor, the company is seen as a threat.

3 *Functional/generic benchmarking*. This is a comparison of specific processes with 'best in class' in different industries, often considered to be world-class in their own right. 'Functional' relates to the functional similarities of organisations, while 'generic' looks at the broader similarities of businesses, usually in disparate operations. With functional benchmarking the partners will usually share common characteristics in the industry, whereas generic benchmarking is not restricted to an industry. It is usually not difficult to obtain access to other organisations to perform this type of benchmarking. Organisations are often keen to swap and share information in a network or partnership arrangement, particularly when no direct threat is presented to a company's business or market share.

There are a number of steps in a formal benchmarking process. They are now briefly described:

- Identify what is the subject to be benchmarked (e.g. the invoicing process), decide who will be in the team, the support they require (e.g. training and project champion) and their roles and responsibilities, reach agreement on the benchmark measures to be used (e.g. number of invoices per day, per person), create a draft project plan and communicate with the required internal parties.
- Identify which companies will be benchmarked from a set of selection criteria defined from the critical success factors of the project. Research potential partners and select the best partner(s).
- Develop a data-collection plan. Agree on the most appropriate means of collecting the data, the type of data to be collected, who will be involved and a plan of action to obtain the data (e.g. explore benchmarking databases, identify contacts in partnering organisations, the questionnaire(s) to be used and the composition, telephone surveys, site visits, etc.).
- Tabulate and analyse data. Determine the reasons for the current gap (positive or negative) in performance between the company and the best amongst the companies involved in the benchmarking exercise.
- Estimate, over an agreed timeframe, the change in performance of the company and the benchmark company to assess whether the gap is going to grow or decrease, based on the plans and goals of the parties concerned.
- Define and establish the goals to close or increase the gap in performance. This step requires effective communication of the benchmarking exercise.
- Develop action plans to achieve the goals. This step involves gaining acceptance of the plans by all employees likely to be affected by the changes.
- Implement the actions, plans and strategies. This involves effective project planning and management.
- Assess and report the results of the action plans.
- Reassess or recalibrate the benchmark to assess if the actual performance/ improvement is meeting what has been projected. This should be conducted on a regular basis and involves maintaining good links with the benchmarking partners.

Success Factors

The choice of benchmarking partners is critical in the success and failure of the project, so it is important that due care and attention are paid to the selection. When contacting potential benchmarking partners, it is helpful to identify the specific areas of activity and the measurement of success which are to be discussed during the visit. It is often useful to send out a pack of information to those organisations that, in principle, have agreed to participate in a benchmarking project. The following is typical of the information it needed to include:

- Covering letter including the reason for undertaking the benchmarking project.
- Overview of your organisation.
- Details of the process being benchmarked, including KPIs and their definitions and descriptions.
- Benchmarking code of conduct to be signed by both parties; reaching an agreement on this encourages openness and trust between the partners.
- Data-collection plan.
- Questionnaire requesting data from the benchmarking partner and a completed questionnaire by the benchmarking team reflecting the state of the art of the process being benchmarked.

The benchmarking teams rate and select their partners on the basis of their critical success factors in terms of what needed to be achieved by the benchmarking project as well as such aspects as comparable size, structure, geography (where deemed appropriate), reputation with respect to product and service quality and market position and segmentation using a criteria rating form to focus on the critical few. Consideration should also be given to the potential partner's understanding of and experience with benchmarking.

The key findings from each benchmarking visit must be related to an action plan with respect to what was to be or had been implemented. This helps to ensure that the best practices that are identified are captured and acted upon. In addition, it is important that the analysis identifies common themes from the benchmarking visits. Simple graphical displays are useful to communicate to all concerned the comparison of the KPI of the process being benchmarked with those of the partners. This aids the acceptance of the changes that need to be made, and regular communication of progress, which should be built into the project plan after the completion of each phase, keeps everyone concerned informed of progress before being presented with the project findings.

As benchmarking is about breakthrough improvement and the implementation of best practices, looking within the industry is sometimes insufficient, as often the 'best' at particular practices are from diverse areas. This is a common problem for very specific benchmarking projects.

It is important to contact the benchmarking partners early. It can be more difficult than expected with respect to the time involved and the issue of identifying the right partner(s). Desk research into the companies being considered as benchmarking partners should be undertaken before deciding, although this does depend on time constraints and the type of project being tackled. Ideally you should aim to visit four to five organisations, and to make all the visits within a period of one month. However, as the company is looking for high gains in the long term from benchmarking, it is worthwhile taking the time to ensure that the company being benchmarked is suitable for analysis.

After each visit to a benchmarking partner, it is important to detail and collate what has been learnt as quickly as possible while the experience is fresh in the team members' minds. It was also found helpful to summarise what the organisation was doing better than the benchmark partner in a report format, identifying key points and providing quantitative as well as qualitative data.

Considerations When Planning Benchmarking Activities

For a benchmarking project to be a success, there are certain difficulties and pitfalls which must be avoided. Based on the projects undertaken, the most common ones are:

- *Unrealistic assumptions.* When planning the actual project, realistic assumptions need to be made about the time required to complete the individual steps of the benchmarking project, the resources needed and the commitment of employees, other than the team members. The planning of the project needs to be as pragmatic as possible. It is also important to 'manage' the expectations at the senior management level in respect of quick results and instant benefits, as well as their role in the benchmarking process.
- *The team members must be free to participate in the project.* The activities associated with the project should not become something else team members do as part of their normal working week, as this will hamper progress and may seriously affect timescales, commitment, and eventually the findings of the project.
- *Lack of a contingency plan.* If the project plan is based on a single set of circumstances or conditions, it is extremely vulnerable to changes. It is essential that a contingency plan be prepared to support the implementation and prepare for any unexpected major changes to the project. This contingency plan must be developed to cope with both favourable and adverse changes. If implementation is broken down into several sequential steps, then it must be possible to bring phase 2 forward if phase 1 takes less time than was initially expected, just as phase 2 would be delayed if phase 1 took longer than expected.
- *Failure to update the plan.* Too often the creation of a plan is seen as a means to an end. Instead, the plan should be considered as a living document which is based on certain assumptions, such as time, cost, resources and levels of commitment, as well as external factors such as the benchmarking partner's response and the time of year. These assumptions will almost certainly change over the period of a benchmarking project, requiring the plan to be updated in terms of what is required and when. In any reconfiguration of the original project, the assumptions should be considered and any necessary changes made. For example, if the project is to finish on time, more money may have to be spent on resources than was originally estimated to achieve any results of significance.
- *Failure to communicate the plan.* Communication of what has been done, what is currently being done and what is planned is vital to the success of a project. If people are not fully aware of what is expected of them, the type of information which has been gathered about best practice, how the benchmark information is to be used to initiate improvements and the changes that will result from the implementation of best practices found from the benchmarking project, then it is highly likely that the plan will fail. It is also important to consider what needs to be communicated and the details, as well as how it should be done.
- *Inadequate project definition.* If the benchmarking team is not aware of why it is doing a particular project and its capability to change a process, then the project will lack direction and focus, leaving the team unsure of what to measure and what best practices it is looking for.

- *Inadequate process understanding*. When documenting a process which is being benchmarked, it is important that not only the processes be described but also each process step, plus the main practices. When carrying this out, the question 'How do we know this?' should be asked several times (i.e. the 5 'whys' approach) to validate what the team considers to be the process with those who are involved at each step. If this is not done, then any conclusions drawn from the benchmarking study may be invalid and a potential danger to the present process.
- *Team members try to do everything themselves*. It is important that the team members do not become insular and try to do everything in relation to the project by themselves. At times they will need to seek the advice and help of individuals who are not directly involved in the benchmarking project. This assistance may be in areas such as data collection, where the data required is already being collected by someone either within a department or externally.
- *The subject area is too large*. Unless the process is within the control of the team and within its comprehension, then it is very difficult to both measure in meaningful terms what is done and ask the right questions of the benchmarking and business partners (e.g. customers and suppliers).
- *It seems like a good idea to use benchmarking* (i.e. it is the latest fad and fashion). Benchmarking, just like any other quality management technique, when used inappropriately, will not bring the expected benefit to the business. Therefore, a balance should be reached between the scope of the problem, the return on investment which is expected and the level of improvement. There is little point in spending considerable time, money and resources on benchmarking a process which will not affect customers in any significant way by bringing breakthrough improvement to business operations.

Advantages of Benchmarking

Deep and valuable knowledge and learning can be gained from both the process and the results of benchmarking (Dehe et al., 2024). There are numerous reasons for an organisation to adopt benchmarking, these include:

- It aids strategic decision-making as your organisation will have a better understanding of how they compare to their competitors.
- It exposes the organisation to new information, opportunities and ways of working.
- Each benchmarking activity can be customised to fit the needs of the business at that particular time.
- It should result in an action plan for improvement.

Disadvantages of Benchmarking

A number of considerations have already been mentioned above; however, there are some common disadvantages:

- Lack of context, benchmarking data can focus on what has been achieved without identifying how it was achieved.
- Lack of consistency in measurement. Different organisations have different definitions for metrics and measure them in different ways, which makes comparisons more difficult.

- Identifying benchmarking partners can be difficult and time-consuming.
- Benchmarking only tells you what has happened, not what the benchmarking partner will do in the future.

Summary of Benchmarking

Benchmarking is a technique for the continuous improvement of processes. It is therefore important to ensure that the process of benchmarking is thought of in a similar vein; the objective is the continual improvement of the benchmarking process used for each project by sharing each project's successes, pitfalls, and failures and thereby promoting continuous learning. There is also a need to ensure that benchmarking is incorporated into an organisation's culture of continuous improvement. A benchmarking project is likely to generate additional benchmarking projects within the process studied or with interfacing processes. A project, in addition to the savings generated, is also helpful in promoting understanding of key performance indicators and measures of quality; in other words, what do we need to have in place to understand what we do, how we do it, why we do it, and how well we do it?

Value Stream Mapping

Value Stream Mapping (VSM) builds on process maps and flowcharts to provide a fact-based process description for understanding the current problems and thinking about the future states. It is a powerful tool that allows the team to communicate and assess how the process should work and perform once the waste and non-added-value activities have been removed, and as such, it is an excellent vehicle for involvement and participation.

The application of VSM offers many advantages for an organisation, it is useful to:

- Promote process thinking in an organisation
- Minimise waste in processes
- Facilitate continuous improvement

How to use Value Stream Mapping

At its core VSM is an elaborated process map which includes data on work in progress, set up time, processing time, error rates, idle time, etc, as well as the information regarding the flows. It is a fundamental technique and as such can form the basis of a Lean, Six-sigma, QM or BRP approach (Gurumurthy & Kodaly, 2011; Winkel et al., 2015). The improvement team must decide on the appropriate level and on the boundaries of the VSM. A high-level perspective is often recommended at the start in order to depict the major elements and their interactions before drilling down into increasing levels of detail where opportunities for improvement are identified. An understanding and analysis of the big picture (Production, Human Resources, Marketing, Finance, Engineering, etc) will allow you to determine the strategic importance of the improvement opportunities. However, a low-level analysis, which includes specific value-added and non-value-added activities, will be essential to generate the breakthrough improvement which is required.

It is usual to map the current existing process first and then the ideal, future state process. This should encourage the team to think outside of the constraints and think innovatively by stretching their imagination. There are a number of common steps involved in producing a VSM:

1 Determine the process, product, or service to be mapped
2 Draw the process flow
3 Add the material flow
4 Add the information flow
5 Collect the process data and add it to the VSM
6 Add process and lead time data to the VSM
7 Verify and validate the VSM

Following these steps will allow you to produce a VSM, an example of which is given in Figure 8.4.

These are agreed and defined; then communicated using a 'Future State' map (See Figure 8.5).

Once the Current State VSM has been produced, then the improvement team should develop a Future State VSM (see Figure 8.5) which describes the process as it should be carried out in the future. This will entail the use of several creative thinking techniques and often includes using the 'clean slate' approach that was discussed in the section BPR. Input from diverse stakeholders (particularly the customers of the process) will be useful to build the Future State VSM to ensure that their needs are being met by the proposed improvements, identify potential sources of waste and non-value-added activities. In this case the concept of value must be related to the customers of the process; if the customers of the process do not receive any benefits for an activity in the process, then it can be said that it is non-value adding.

Once the Future State VSM has been created, it is useful to test it where possible through simulations or piloting to test if it will deliver the expected results before widespread implementation throughout the organisation.

The process of improvement utilising VSM can be summarised as follows:

1 Organising the improvement team and prioritising their workload to facilitate the organisation's strategy
2 Pre-mapping the selected processes (process mapping)
3 Developing the basic VSM
4 Collecting the process data and engaging with the different stakeholders
5 Identifying the current state (producing the Current State VSM)
6 Visualising the future state (producing the Future State VSM)
7 Running some simulation and collecting performance data
8 Implementing the changes
9 Reviewing the improvement

The core application of VSM involves creating a 'Current State' map (see Figure 8.4) then identifying the necessary improvements.

Advantages of Value Stream Mapping

VSM provides the organisation with a number of advantages:

- It is a systematic methodology for process improvement
- It develops a culture of waste removal

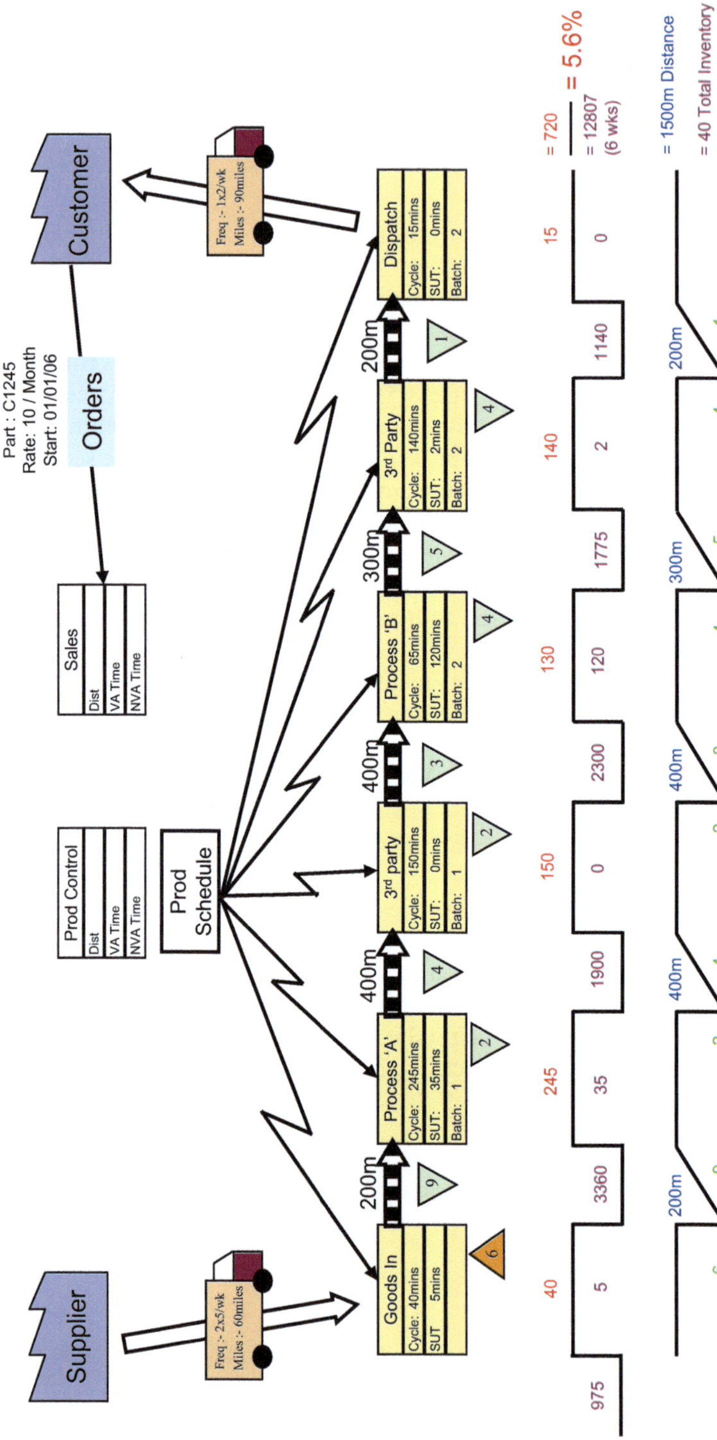

Figure 8.4 VSM current state map.

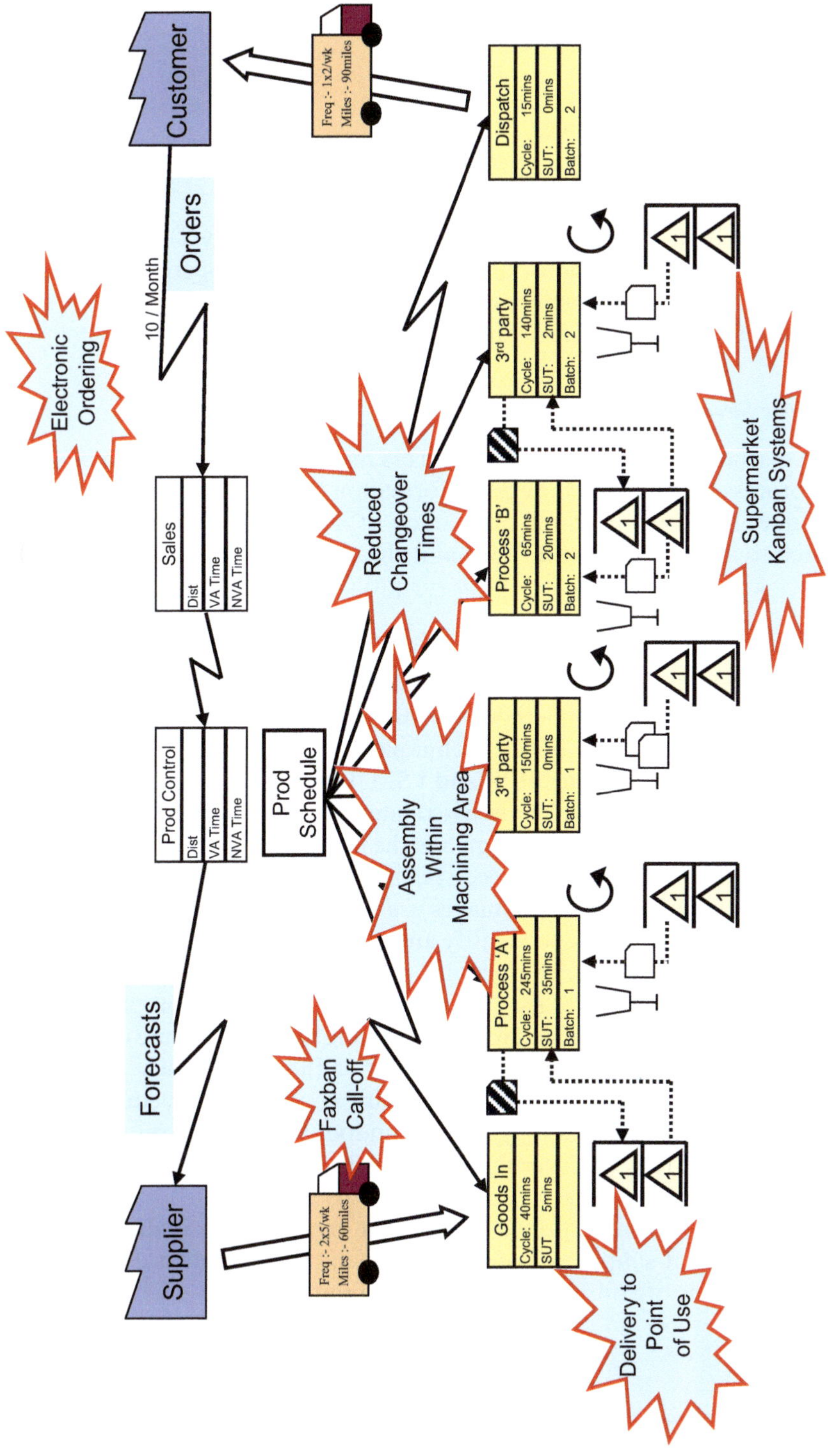

Figure 8.5 VSM future state map.

- The concept of non-value-added activity is particularly important for service organisations
- It encourages creativity in process redesign.

Disadvantages of Value Stream Mapping

VSM does have a number of drawbacks:

- It works best for high-volume, low-variety operations. It can be cumbersome and inappropriate to try to implement it for high-variety, low-volume operations
- The concept of value can be subjective and difficult to measure
- It does not consider the physical layout of the process sufficiently

Summary of Value Stream Mapping

A VSM is, in essence, a specialist form of process map. It is considered to be more useful than regular process maps due to the extra information that it presents. The real value to an organisation does not come from the production of a VSM, but the adoption of a structured approach to improvement with VSM at its heart as described above.

Six-Sigma

Six-sigma is a data-driven integrated process improvement and problem-solving approach that utilises statistical tools and methods (Widiwati, Liman, & Nurprihatin, 2025), and encompasses and is built on all of the techniques described previously in this chapter: QFD, DoE, FMEA, Benchmarking, BPR and VSM. It has popularised the use of Statistical Process Control (SPC), which will be discussed later in this section.

Motorola created the concept of six-sigma in the mid-1980s to improve the performance, productivity and quality of their key processes, partially in response to demands from customers for improved quality. Six-sigma makes use of quality engineering methods within a defined problem-solving structure to identify and eliminate process defects and solve problems, and in this improve output, productivity, operating effectiveness and customer satisfaction (Gijo et al., 2014; Jesus et al., 2015). Like SPC it focuses on understanding and eliminating the causes of variation and robust designing for manufacture or service delivery.

Many of the objectives of six-sigma are similar to those of quality management (e.g. customer orientation and focus, team-based activity, comprehensive education and training, and problem-solving methodology), and it undoubtedly builds on these. There is no doubt that six-sigma brings engineering and statistical analysis back into quality and is returning quality back to its roots.

It seems that six-sigma is interpreted by different organisations in different ways. Some organisations interpret it simply as a measurement and improvement device, while others use it as a label for their organisation-wide quality approach.

What Is Six-Sigma?

A sigma is a statistical measure of variation in terms of the standard deviation of the variable being analysed. It indicates the spread of each measurement around the target value, and therefore, it is essentially an indication of how good a product or service is. Traditionally, designers used the three-sigma rule to evaluate whether or not a part would meet specifications. When a part's specifications are consistent with a spread of six standard deviations of process variation (three sigma to either side of the target value), around 99.73 per cent of the parts would be expected to conform to specification. In essence, sigma measures the capability of a process to produce defect-free work and is a means of calibrating process performance to meet the requirements of customers. For example, a process that is at a quality level of three sigma means 66,807 defects per million opportunities (DPMO), while six-sigma is 3.4 DPMO. Other sigma levels and their corresponding number of defects are presented in Table 8.1.

Or, presented differently, this can be represented as per Figure 8.6.

Basic statistical quality tools and techniques, such as SPC, are seen as the cornerstone of six-sigma. Often, these tools and techniques are found to already be in place before the companies start implementing six-sigma, which indicates that six-sigma is usually based on existing knowledge and practices.

Six-sigma is commonly implemented to achieve the following:

- Reduce the number of defects
- Reduce variation
- Increase quality

Table 8.1 Six-sigma and defects per million opportunities

Sigma	*Defects per million opportunities (DPMO)*
2	308,537
3	66,807
4	6,210
5	233
6	3.4

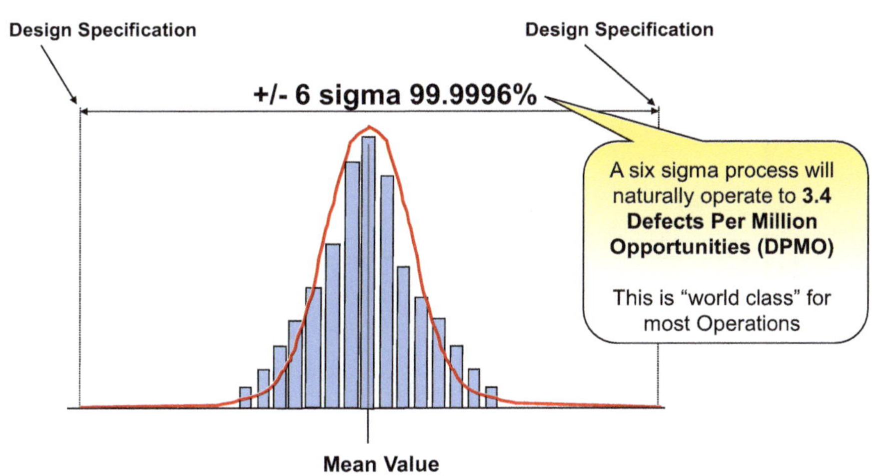

Figure 8.6 Six-sigma and defects per million opportunities.

- Improve customer satisfaction
- Improve profitability

Statistical Process Control

Statistical process control (SPC), although popularised by the use of six-sigma is not a new concept; its roots can be traced back to the work of Shewhart (1931) at Bell Laboratories in 1923. SPC is about control, capability and improvement, but only if used correctly and in a working environment which is conducive to the pursuit of continuous improvement.

On its own SPC will not solve problems; the control charts only record the 'voice of the process' and simply confirm the presence of a problem. This is why it is often associated with six-sigma improvement initiatives. The application of SPC can potentially be extensive. It is not simply for use in high-volume manufacturing or processing; it is equally applicable to service processes and procedures.

An important means of improvement is the reduction of variation. SPC is very useful for six-sigma initiatives because it identifies the extent of the variation and whether it is due to special or common causes, process improvement being achieved by removing either or both causes. It should be stressed that while SPC can indicate the amount of variation, it will not give the source. The techniques described elsewhere in this section can be used to identify the source of the variation and minimise or eliminate it.

There are four main uses of SPC:

1 To achieve process stability.
2 To provide guidance and understanding on how the process may be improved by the reduction of variation and to keep it reduced.
3 To assess the performance of a process.
4 To provide information to assist with management decision-making.

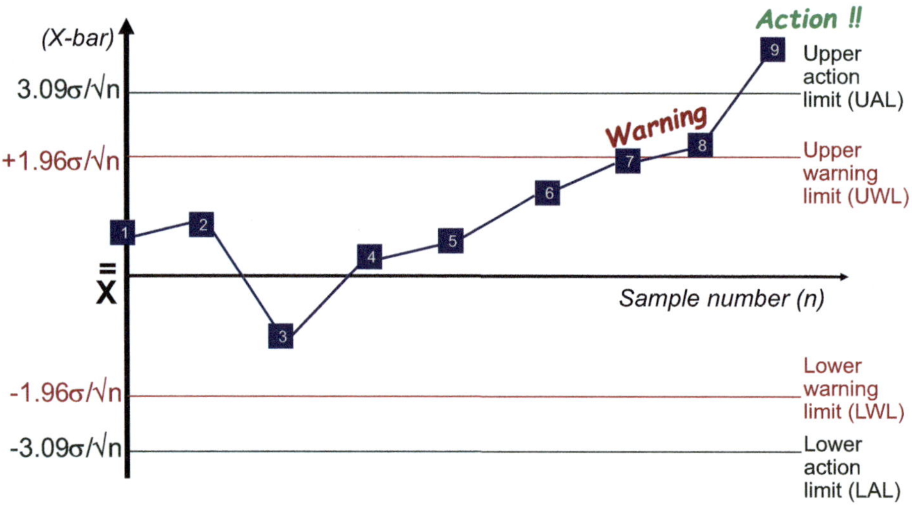

Figure 8.7 Sample SPC chart.

How to Use Statistical Process Control

The first step in the use of SPC is to collect data according to a plan and plot the gathered data on a graph called a control chart, for example, see Figure 8.7. Control charts are a useful tool to visualise the level of variation of a process to identify the sigma level that is currently capable of operating at. The next task is to reduce, as much as possible, the common causes of variation so that the output from the process is centred around a nominal or target value. This is an iterative process in the pursuit of continuous improvement.

What are Special and Common Causes of Variation?

Special (or *assignable*) *causes* of variation occur intermittently in the form of shocks and disturbances to the system and reveal themselves as unusual patterns of variation on a control chart. Special causes should be identified and rectified, and hopefully, with improved processes or even product design, their occurrence will be minimised. In the short term, their presence should be highlighted and a response programme established to deal with them.

Common (or *unassignable*) *causes* produce the natural or random pattern of variation observed in data when they are free of special causes. Common causes arise from many sources and do not reveal themselves as unique patterns of variation; consequently, they are often difficult to identify. If only common cause variation is present, the process is stable and predictable.

In the pursuit of process improvement, it is important that a distinction is made between special and common cause sources of variation because their removal may call for different actions. Cross-functional teams are often set up to eliminate special and common causes of variation. SPC is widely used in quality improvement to monitor the performance of processes and to assess the effectiveness of quality improvement initiatives.

Figure 8.8 shows the ideal impact of a quality improvement initiative on a process; the upper and lower specification limits have narrowed, which displays a reduction in common cause variation.

Advantages of Statistical Process Control

There are numerous advantages to employing SPC as a quality control and improvement tool:

- The charts are easy to read and to form a consensus about.
- It facilitates a focus on the removal of variation.
- The concept of common cause and special cause variation allows different approaches to be used for different causes.
- The charts quickly indicate if a change has made an improvement (or made things worse!).
- Software is available to aid and automate the creation of the charts.

Difficulties Experienced in Introducing and Applying SPC

A number of difficulties with the introduction of SPC have been identified:

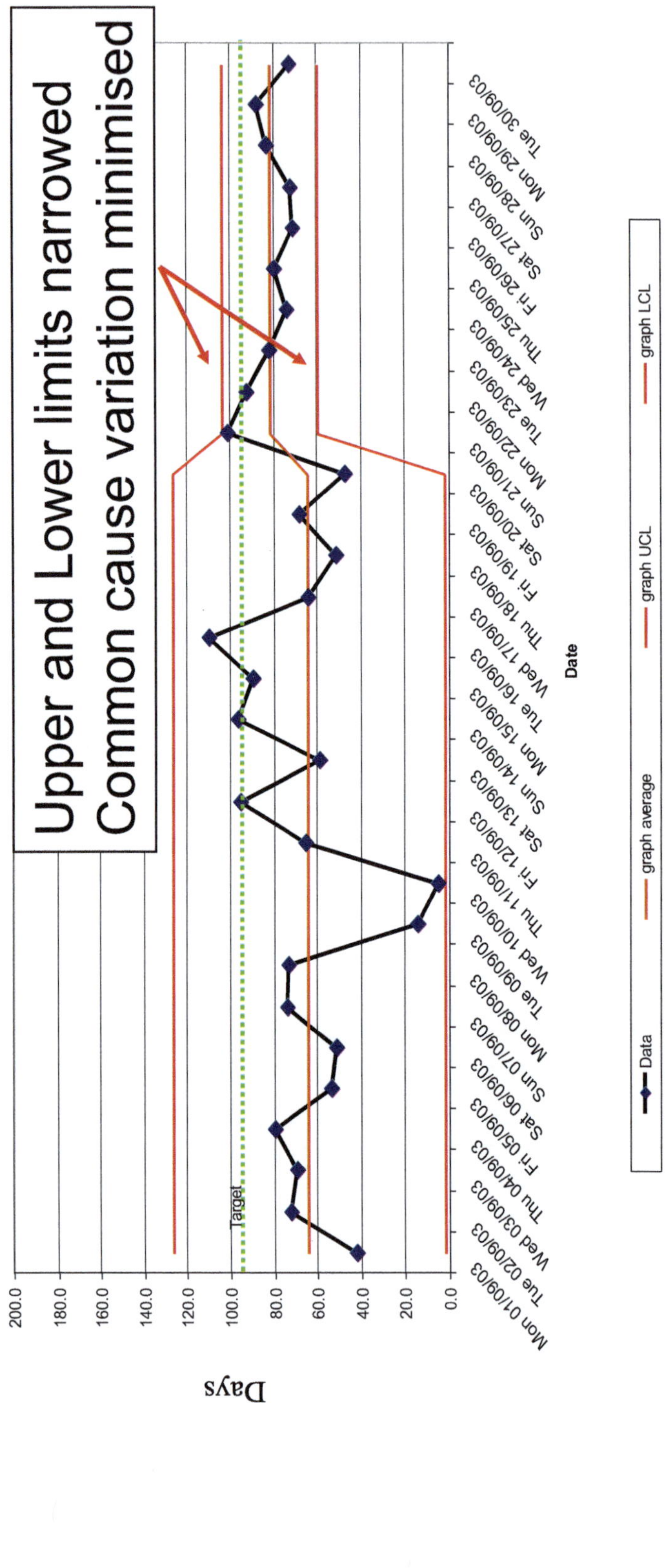

Figure 8.8 SPC chart after limit change.

- It requires good-quality data
- Lack of knowledge of/expertise in SPC, especially amongst senior management
- Poor understanding and awareness within the company of the purpose of SPC
- Requires specialist process knowledge to identify the causes of the variation.

Summary of SPC

SPC has proved to be a very useful measurement technique to identify the amount of variation that can be expected of a process. As such, it supports six-sigma initiatives as well as a very visual tool to identify variation and monitor improvements and processes.

How to Use Six-sigma

The implementation of six-sigma, like any major organisational change programme, is not easy. Its success will depend on at least four major factors.

Firstly, it involves high levels of commitment and involvement from senior management. It is based on an understanding of statistics, and this is not a popular area for most managers. It also requires that high-performing managers are released to be trained and, after training, that they commit a significant amount of their time to the six-sigma concept.

Secondly, it cannot be treated as yet another stand-alone activity. Like QM, it requires adherence to a whole philosophy rather than the usage of a few tools and techniques, however sophisticated. The organisation must be used to working with cross-functional teams and should have its major processes identified and under some degree of control. In short, it must have many of the fundamentals of QM already in place.

Thirdly, six-sigma is about reducing defects. Improvement depends on how opportunities for defects or failure are defined and measured. What matters most is what the customer wants or needs; this is why QFD (see above) is also a technique strongly associated with six-sigma. The most critical effects or failures are those that most concern the customer.

Fourthly, six-sigma needs to be focused on those parts of a business which will influence customers to prefer dealing with this company over one of its competitors. Therefore, six-sigma, like any performance improvement drive, should start from strategy – where does a company want to be? What will really make a difference to getting there? And, therefore, where must it concentrate its improvement drives?

Successful six-sigma implementation needs to build on the experience of earlier quality management programmes. Having developed quality awareness and a quality culture and having reached a certain level of quality management maturity are essential prerequisites for the success of six-sigma. There will always be potential obstacles to sustainable six-sigma implementation: the benefits of projects may diminish over time, management's focus may shift to other priorities, and important players in the organisation may lose interest.

The Core Elements of Six-sigma

Six-sigma builds on a range of improvement methods that have proven to be effective. This can be seen in its central themes:

- *Focus on the customer*. Six-sigma measures start with customer satisfaction. The emphasis is on understanding customer expectations and requirements.
- *Data- and fact-driven management*. This is a classical quality management theme, including speaking with data, management decisions based on fact, and developing an in-depth understanding of internal processes.
- *Specific training*. A defined and formal infrastructure of champions, master black belts, black belts and green belts that head and influence six-sigma projects. Master black belts are the technical experts who provide training and support for the other belts; they also lead major cross-functional six-sigma projects. The black belts undertake full-time work on six-sigma projects and lead the project teams. The green belts are part-time process owners and usually undertake work on a small scale.
- *Structured approach*. Six-sigma is based on a structured problem-solving approach. For existing processes, the approach consists of the following steps: define, measure, analyse, improve, and control (DMAIC); for new processes the steps are define, measure, analyse, design, and verify (DMADV). Both approaches are discussed below.
- *Quality engineering*. Six-sigma uses a full range of tools and techniques. A number of writers on six-sigma suggest the application of specific tools and techniques against each stage of the DMAIC and DMADV problem-solving approaches.
- *Process focus, control and improvement*. The key aspect is understanding the process to control its input and thereby facilitate its improvement. This involves an examination of potential defects, root causes and potential corrective and long-term actions. It is important to understand the relationship between inputs and outputs.
- *Proactive management*. Management at all levels must attempt to understand the key principles of six-sigma. They must be active in challenging why things are done in a certain way, defining root causes of problems, setting and maintaining aggressive improvement targets, and being prepared to devote a large amount of their time.
- *Boundaryless collaboration*. Teamwork is an essential part of six-sigma and it is important to have a range of skills within the team.
- *Drive for perfection*. In the drive to eliminate defects it is important to accept that from time to time things will go wrong, and some projects will not achieve their goals. It is important to understand the reasons for failure, to learn from experiments, and to put in place countermeasures to prevent defects from occurring in the future.
- *Cost savings of each project*. A sense of urgency is created with six-sigma, through the financial targets linked to each project. Each six-sigma project should lead to verifiable bottom-line results.
- *Short-term improvement projects*. A key requirement is that a timescale is agreed for a specific project's completion. Moreover, the duration of each project is relatively short, with a typical project lasting between three and six months.

Structured Problem-Solving Approaches

Six-sigma improvement projects adhere to strict problem-solving approaches. Depending on the organisational processes to which they are applied (i.e. existing processes or new processes), improvement projects use either the DMAIC or DMADV approach. Implementation of six-sigma involves three aspects of process development: Process improvement; Process design/redesign; Process management.

Process Improvement – DMAIC

This primarily concerns the elimination of the root causes of process problems and is clearly associated with continuous improvement activities (i.e. improve what you already have). Most six-sigma activities are initially based on process improvement. It involves the identification of the vital few issues that influence results. The DMAIC structured problem-solving approach, which is employed in making improvements to existing processes and products, is related to other problem-solving approaches, including Deming's PDCA cycle. The DMAIC approach takes the following steps:

- *Define* in clear terms the specific problem to be worked on and the process to be improved. The problem needs to be one where it is critical to succeed and/or one that will give the quickest or greatest returns. This phase also involves defining the project scope and boundary conditions, selecting appropriate performance metrics, and agreeing on the goals of the selected project, its financial impact, and the project champion, process owner and team members.
- *Measure* the factors which are critical to quality. This involves selecting the process outputs to be improved, developing a data-collection plan, gathering data to evaluate current performance and assessing the performance measurement systems and their capability.
- *Analyse* the relationship between the cost of defects and key process variables. The purpose of this is to determine the root cause of variation and defects.
- *Improve* the process using experimentation, pilot studies and simulation techniques to address the root causes of the problems identified in the analysis phase. This involves various improvement loops, with appropriate confirmation studies related to each phase of process improvement.
- *Control* the process outputs to ensure the long-term gains and improved performance of the process. This involves verifying the benefits and savings, ensuring that the changes are fully integrated into procedures, and communicating the findings as appropriate.

Process Design/Redesign – DMADV

This is normally the second stage after process improvement. The emphasis is on new processes rather than fixes to existing ones. The method used is the DMADV approach, also called design for six-sigma (DFSS). The steps are different from those of the DMAIC method, and can be described as:

- *Define* the project goals and customer (internal and external) deliverables
- *Measure* and determine customer needs and specifications
- *Analyse* the process options to meet the customer needs
- *Design* (detailed) the process to meet the customer needs
- *Verify* the design performance and ability to meet customer needs

Process Management

This third phase reflects a change of emphasis from oversight and direction of functions to understanding and process facilitation. The key activities include: processes are managed end-to-end; customer requirements are clearly identified; meaningful measures of input, process activities, and output are developed.

A six-sigma approach can be considered to have the following five stages:

- *Stage 1 – Identify the need.* Gather data; analyse data; define cost of poor quality; consider the voice of the customer.
- *Stage 2 – Clarify the vision.* Define project brief; set goals/target objectives; establish realistic timescales.
- *Stage 3 – Develop the plan.* Establish the project initiative; identify the project champion; select the team; understand the present documentation system; review the current process; understand current process performance; consider training requirements; define the available resources.
- *Stage 4 – Implement the plan.* Promote clear process ownership; optimise the process; record progress results and display using appropriate visual management methods; evaluate improvement and process management; review cost savings; establish continuous improvement mechanisms.
- *Stage 5 – Sustain improvement.* Document the initiative; maintain team responsibility; apply knowledge gained across the board; communicate and recognise success; identify the next project.

Advantages of Six-Sigma

Implementing six-sigma in your organisation will deliver:

- Process optimisation
- Data-driven decision-making
- Improved quality
- Improved customer satisfaction

Disadvantages of Six-Sigma

There are a number of drawbacks to embedding six-sigma:

- The requirement for training a large number of staff
- The time taken to fully implement six-sigma across the organisation
- The requirement for good-quality data
- It can build in inflexibility and an over-reliance on certain experts

Summary of Six-Sigma

Six-sigma is a well-structured improvement approach with verifiable financial results. However, organisations' quality initiatives should not be driven solely by the savings needed to impress the financial analysts. This can be a dangerous approach because its focus is too short-term. Most organisations need a longer-term focus on customers, and not solely on suppliers of capital. A successful six-sigma implementation should build upon several prerequisites, such as an existing quality culture and a certain level of quality maturity. The sustainability of six-sigma in the long term depends on many factors, such as top management commitment, being able to show successful projects, high investment in training, high investment in management time, and the involvement of key players in the organisation.

Six-sigma can revolutionise an organisation, and it will go deep into its fabric; it therefore needs top management drive behind it. It must be seen as part of a total approach, and it demands a level of quality competence from the organisation before the benefits can begin to be delivered. Quality improvement methods such as six-sigma may be very powerful, but they must be directed and need a clear strategy to measure and interpret customers' needs successfully. Managers must also realise that, if their organisation has no clear power structure and the desired level of competence is not present, then a six-sigma programme is unlikely to work.

References

Arboretti, R., Ceccato, R., Pegoraro, L., & Salmaso, L. (2022). Design of Experiments and machine learning for product innovation: A systematic literature review. *Quality and Reliability Engineering International*, 38(2), 1131–1156.

Bamford, D., Forrester, P., & Reid, I. (2023) *Essential guide to operations management: Concepts & case notes*, 2nd edition. UK: Routledge.

Dale, B, Bamford, D, van der Wiele, T, (2016) *Managing Quality: An essential guide and resource gateway*, 6th edition (p. 330). UK: John Wiley & Sons.

Dehe, B, & Bamford, D. (2017). Quality Function Deployment and operational design decisions: A healthcare infrastructure development case study. *Production Planning & Control*, 28(14), 1177–1192.

Dehe, B., Bamford, D., & Kotcharin, S. (2024). Bespoke benchmarking framework employed as vehicle and platform for open innovation: A healthcare infrastructure case. *Production Planning and Control*, 35(7), 725–743.

Gijo, E.V., Bhat, S., & Jnanesh, N.A. (2014). Application of Six Sigma methodology in a small-scale foundry industry. *International Journal of Lean Six Sigma*, 5(2), 193–211.

Gurumurthy, A., & Kodaly, R. (2011). Design of lean manufacturing systems using value stream mapping with simulation: A case study. *Journal of Manufacturing Technology Management*, 22(4), 444–473.

Hariri, A., Domingues, P., & Sampaio, P. (2023). Integration of multi-criteria decision-making approaches adapted for quality function deployment: An analytical literature review and future research agenda. *International Journal of Quality & Reliability Management*, 40(10), 2326–2350.

Jesus, A.R., Antony, J., Lepikson, H.A., & Teixeira Cavalcante, C.A. (2015). Key observations from a survey about Six Sigma implementation in Brazil. *International Journal of Productivity and Performance Management*, 64(1), 94–111.

Kassela, K., Papalexi, M., & Bamford, D. (2017). Applying quality function deployment to social housing?. *The TQM Journal*, 29(3), 422–436.

Shewhart, W.A. (1931). *Economic control of quality of manufactured product*. New York: D. Van Nostrand Co. Inc.

Taguchi, G. (1986). *Introduction to quality engineering: Designing quality into products and processes*. Tokyo: Asian Productivity Organisation.

Twaissi, N.M., Al Zoubi, M.O., Masa'deh, R.E., Zoubi, J.Z.A., & Alhelalat, J.A. (2025). The mediating role of TQM interventions and benchmarking in the impact of organizational culture on organizational performance. *Journal of Industrial Engineering and Management*, 18(1), 10–32.

Widiwati, I.T., Liman, S.D., & Nurprihatin, F. (2025). The implementation of Lean Six Sigma approach to minimize waste at a food manufacturing industry. *Journal of Engineering Research*, 1;13(2), 611–626.

Winkel, J., Edwards, K., Birgisdóttir, B.D., & Gunnarsdóttir, S. (2015). Facilitating and inhibiting factors in change processes based on the lean tool 'value stream mapping': An exploratory case study at hospital wards. *International Journal of Human Factors and Ergonomics*, 3(3–4), 291–302.

9 Quality Management Systems

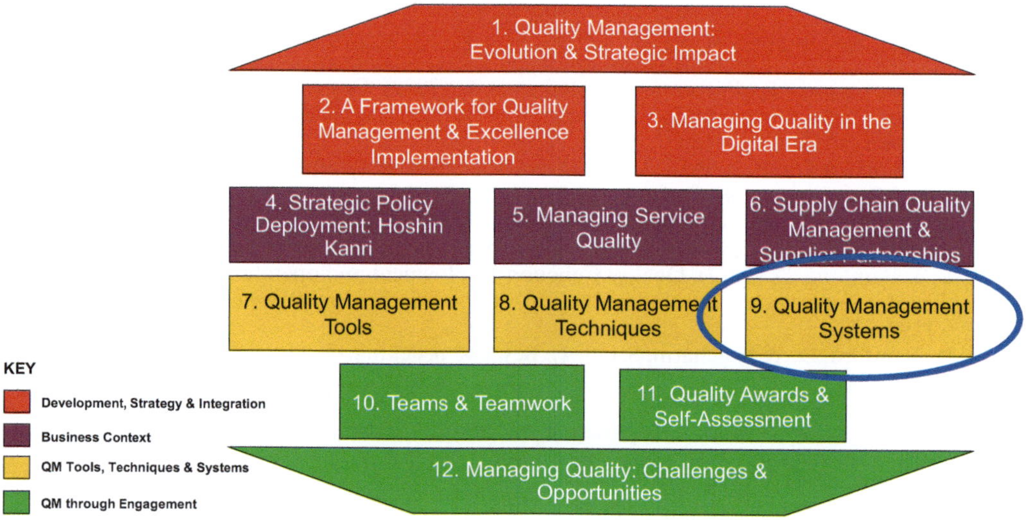

Introduction

This chapter opens by examining the concept of quality assurance and the responsibilities of people within an organisation for carrying out the activity. Quality Management Systems (QMS) are defined, the background of quality system standards traced, the key features of QMS are discussed and illustrated using the ISO 9000 series (2015), implementation guidelines and issues are outlined, quality system assessment and registration are reviewed, and the benefits and limitations are highlighted. We have chosen to focus on the ISO Standards as these have been designed to be broadly applicable to and sizes of organisation, in any industry (Bamford, Forrester, & Reid, 2023), and in the past few years an increasing number of countries that want to compete at an international level have adopted this standard (Sherwani, Demir, & Maroof, 2025). A model is also presented, which outlines what is required for a small company to successfully achieve QMS registration. Much has already been written about quality systems and standards in addition to the standards themselves. This chapter is therefore restricted to an overview of the key features and issues.

DOI: 10.4324/9781003594369-12

What Is Quality Assurance?

Quality assurance is defined by ISO 9000: 2015 as:

> A part of quality management, which is focused on providing confidence that quality requirements will be fulfilled.

Quality assurance is often regarded as discreet policing by the quality assurance department. This is not so. The ideal role of the department is to oversee the whole process of quality assurance within an organisation, provide guidance and advice on the assignment of roles and responsibilities to be undertaken by each function and person, and address weaknesses in the system. Quality assurance needs to be an integral part of all an organisation's processes and functions, from the conception of an idea and throughout the life cycle of the product or service: determining customer needs and requirements, planning and designing, production, delivery and after-sales service.

The objective should be to get every person in the organisation to take personal responsibility for the quality of the processes for which they are accountable. This includes treating the following processes as 'customers' and endeavouring to transfer conforming products, services, materials and documents to them, monitoring quality performance, analysing non-conformance data, taking both short- and long-term action to prevent the repetition of mistakes, and promoting the feed forward and feedback of data. The emphasis should be on the pursuance of corrective and preventative actions and procedures and non-conformance investigation in a thorough manner with closed-loop effectiveness. It is also necessary for everyone to perform their tasks in accordance with their training and procedures, as defined by the quality management system.

The main objective of quality assurance activity is to build quality into the product and/or service during the upstream design and planning processes, and in this way give confidence to a customer that a product and/or service performs as they expect. Quality function deployment, FMEA, design of experiments, design for manufacturability/assembly and quality audits are all part of an advanced product quality planning process, and of considerable assistance in the pursuance of this goal.

Quality assurance activity, which is planned and managed along these lines, will strengthen an organisation's QM and strategic improvement process efforts.

What Is a Quality Management System?

A quality management system is defined by ISO 9000:2015 as:

> a formal set of policies, processes and procedures established to elevate consumer satisfaction.

The purpose of a QMS is to establish a framework of reference points to ensure that every time a process is performed, the same information, methods, skills and controls are used and applied in a consistent manner. In this way, it helps to define clear requirements, communicate policies and procedures, monitor how work is performed and improve teamwork.

Documentary evidence about the quality management system is fundamental to quality assurance and takes several forms.

- A company quality manual (sometimes called a level 1 document) provides a concise statement of the quality policy and quality management objectives as part of the company objectives. ISO 10013:2021 provides useful guidelines on the development and preparation of quality manuals.
- A procedures manual (sometimes referred to as a level 2 document) describes how the system functions, gives the structure and responsibilities for each department/unit and details the practices to be followed in the organisation.
- Work instructions, specifications, methods of performance and detailed methods for performing work activities form a third level of documents.
- In addition, there is often a database containing all other reference documents (e.g. forms, standards, drawings, reference information, and supplier list).

The quality management system documentation helps to ensure that employees know what they should be doing, along with the appropriate means for doing it. It also provides evidence to those who wish to assess the system.

The QMS should define and cover all facets of an organisation's operation, from identifying and meeting the needs and requirements of customers to design, planning, purchasing, manufacturing, packaging, storage, delivery, installation and service, together with all relevant activities carried out within these functions. It deals with organisation, responsibilities, procedures and processes. Put simply, a quality system is good management practice.

A quality management system, if it is to be comprehensive and effective, must cover all these activities and facets and must be developed in relation to the corporate strategy of the company.

A QMS which embraces quality management objectives, policies, organisation and procedures, and which can demonstrate, by assessment, compliance with ISO 9001:2015, provides an effective managerial framework on which to build a company-wide approach to a process of continuous improvement.

The Development of Quality Management System Standards

Regardless of the approach taken to QM or strategic process improvement and the quality management maturity of the organisation, a business may need to demonstrate to customers that its processes are both effective and under control and that there is effective control over procedures and systems. The pressure for proof that systems and procedures are in place and working in an effective manner led to the demand for quality assurance based on the development of quality management system standards. The origins of this can be traced back to the 1950s, when the US Department of Defense and the UK Ministry of Defence saw a need for greater reliability in purchased products and a reduced reliance on customer or purchaser inspection as the main assurances of quality.

In 1972 the British Standards Institution (BSI) published BS 4891, *A Guide to Quality Assurance*, which set out guidance to organisations on quality and its management and was intended as a guide to companies developing their QMS; this standard was withdrawn in 1994. This was followed in 1974 by the issue of BS 5179, which was a three-part standard, *A Guide to the Operation and Evaluation of Quality Assurance Systems*; this standard was withdrawn in 1981 after being superseded in 1979 by the first issue of BS 5750. During the mid-1970s there was a proliferation of quality system standards produced by a variety of second- and third-party organisations. The Warner report (1977), *Standards and Specifications in the Engineering Industries*, stressed the need for a national standard for

QMS, to reduce the number of assessments to which suppliers were being subjected by their customers. It pointed to the shortcomings and fragmented nature of the British system of standards. It was recommended that British Standards be produced to provide a single base document for quality systems. Subsequently, in 1979, the BSI issued the BS 5750 series of quality management system standards.

It was the BSI that formally proposed the formation of a new technical committee (ISO/TC 176) to develop international standards for quality assurance, techniques and practices (this committee is responsible for developing and maintaining the ISO 9000 family of standards). Some 20 countries originally participated in the development of what was to become the ISO 9000 series. In 1987 the series of international standards on QMS was first published by the International Organisation for Standardisation (ISO; a federation of some 140 countries' national standards institutes). This initial version of the standards, whilst reflecting various national approaches and international requirements, was based largely on the 1979 version of the BS 5750 series and the eight or so years of UK user experience, mainly in the manufacturing industry, and the Canadian CSA Z299 series.

The ISO 9000 series was adopted by CEN (the European Committee for Standardisation) and CENELEC (the European Committee for Electro-technical Standardisation) as the EN 29000 series, thus harmonising the approach to quality systems in the European Community, the standard at this stage having three numbers: national, European and international. It has perhaps had the most significant and far-reaching impact on international standardisation of any set of standards. The ISO 9000 series has been revised based on international implementation experience and was reissued in September 2015. The major changes reflect that organisations have evolved into more global and complex supply chain than they did in the past. The ISO 9000:2015 series is based on seven quality management principles (QMPs): Customer focus, Leadership, Engagement of people, Process approach, Improvement, Evidence-based decision making and Relationship management.

Quality Management System Standards: ISO Series

Registration to ISO 9001 is a useful foundation leading to the development of a quality system to meet the independent system requirements of customers. Several major purchasers use this registration as the 'first pass' over a supplier's quality system. They will take ISO 9001 as the base and only assess those elements of the system that they believe are particularly sensitive to them as purchasers.

The International Organisation for Standardisation conducts a survey of the number of certificates that have been issued against the ISO series of standards. The survey reports details by regions and country of registration, as well as an industry-sector breakdown by country. The results for 2023 show that 1,476,801 certificates have been awarded across 2,355,301 sites worldwide (The ISO Survey of Management System Standard Certifications, 2023).

Core Principles

The objective of the ISO 9000 series is to give purchasers an assurance that the quality of the products and/or services provided by a supplier meets their requirements. The series of standards defines and sets out a definitive list of features and characteristics that are considered to be present in an organisation's management control system through documented policies, manuals and procedures, which help to ensure that quality is built into a process and is achieved. Amongst other things it ensures that an organisation has a quality policy, that procedures are standardised, that defects are monitored, that corrective and preventative action systems are in place, and that management reviews the system. The aim is systematic quality assurance and control. It is the broad principles of control, in general terms, which are defined in the standards, and not the specific methods by which control can be achieved. This allows the standard to be interpreted and applied in a wide range of situations and environments and allows each organisation to develop its own system and then test it against the standard. However, this leads to criticisms of vagueness.

The series of standards can be used in three ways:

- Provision of guidance to organisations to assist them in developing their quality systems.
- As a purchasing standard (when specified in contracts).
- As an assessment standard to be used by both second-party and third-party organisations.

The ISO family of standards consists of a number of different standards. The most popular being:

- ISO 9001:2015: which sets out the requirements of a quality management system.
- ISO 14001:2015: which is a standard for environmental management quality systems, providing a framework for organisations to design and implement an environmental quality management system, and continually improve their environmental performance.
- ISO 45001:2018: which is a standard for an occupational health and safety quality management system. It provides a framework for organisations to manage risks and improve occupational health and safety performance.

The standards have two main functions. The first function identifies the aspects to be covered by an organisation's quality system and gives guidance on quality management and application of the standards. The second function defines in detail the features and characteristics of a quality management system, which are considered essential for the purpose of quality assurance in contractual situations.

ISO 9000 outlines the fundamentals of QMS and provides the definitions of the key terms used in the other standards. It presents quality management system requirements applicable to all organisations' products and services. It is used for demonstrating system compliance to customers, for certification of QMS, and as the basis for contractual requirements. It requires the following:

- A detailed documentation of quality requirements, processing steps and results.
- Implementation of a set of controls to maintain the system.
- Compliance with the requirements of the 22 sub-elements.

ISO 9004:2018 is a quality management system guidance specification that embraces a holistic approach to performance improvement and customer satisfaction. Both ISO 9001:2015 and ISO 9004:2018 are based on a process model that uses the following seven QMPs that reflect best practice:

- QMP 1 – Customer focus
- QMP 2 – Leadership

- QMP 3 – Engagement of people
- QMP 4 – Process approach
- QMP 5 – Improvement
- QMP 6 – Evidence-based decision making
- QMP 7 – Relationship management

These two standards employ common vocabulary and structure to facilitate their use and are intended to be used together by organisations wishing to develop their systems beyond the minimum requirements of ISO 9001.

ISO 19011 (2018) will provide guidance on managing and conducting environmental and quality activities. This standard combines the quality system auditing standard (ISO 10011: Parts 1 to 3) with the environmental system audit standards (ISO 14010, ISO 14011, and ISO 14012).

Implementation Guidelines

Organisations that are considering adopting a QMS have the benefit of learning from those organisations that have previously gone through the process of attaining the standards. Whilst this section focuses on the implementation of the ISO series, the principles discussed here can be applied to the implementation of any other Quality Management Standards.

- An organisation should be clear on the reasons for seeking ISO 9001. Implementation for the wrong reasons will prevent the organisation from receiving the full benefits. In addition, it may be found that implementing and maintaining the requirements of the chosen standard is a burden in terms of costs and extra paperwork with no compensating benefits. ISO 9001 registration must therefore not be sought just to satisfy the contractual requirements of major customers or for marketing purposes. Indeed, when most competitors have ISO 9001 registration, there is little marketing advantage, and in many markets, it is now a qualifying criterion.
- The development of a quality management system to meet the requirements of ISO 9001 should be designed and delivered as a project, with the identification of key steps, milestones and timescales. This will prevent progress from being sporadic and variable.
- Prior to the implementation of ISO 9001 it is important that an audit of the existing quality management system is carried out by a qualified auditor. This will provide a gap analysis, which, by determining the initial status of the company's quality management system, will enable management to assess the amount of work required to meet its requirements and to plan the project to systematically implement the standard. Involvement of the appointed management representative during the quality audit is essential.
- For organisations developing a quality management system for the first time, a steering committee should be established comprising all the heads of departments and chaired by the CEO. This type of representation is essential to gain cross-functional support for the project and to help ensure the smooth development and implementation of the system. Participation and commitment from all the heads of department are critical to gain employee support for the project, and this will help to ensure the smooth implementation and subsequent maintenance of the standard. In extremely small companies where there is little or no second-tier management the whole-hearted commitment and involvement of the CEO is critical and essential.

Enhancement of the Quality System

ISO 9001 should be considered as the minimum requirement. Without a documented quality management system, there will be no reliable baseline quality data to initiate and monitor the process of quality improvement. Organisations should, however, aim to have a quality system which surpasses the standard's requirements, with new quality initiatives built into the system, which is the objective of ISO 9004. A quality management system that meets the requirements of ISO 9001 should in no sense be regarded by senior management as the pinnacle of their quality management achievements. All it says to the outside world is that the organisation has controls, procedures and disciplines in place. The organisation should treat ISO 9001 registration as a precursor to developing its approach to QM and strategic process improvement.

There is a need to create a conducive environment for the development of a quality management system, which meets the requirements of ISO 9001. This can be achieved by the formulation of organisational quality policy and quality objectives. The responsibility of executives in the establishment, maintenance and development of a QM system cannot be overemphasised. The leadership of senior management and their total commitment to the process of quality system registration to ISO 9001 are vital. Whilst the CEO has ultimate responsibility, senior management must not only understand the principles of the ISO 9000 series but should ensure that the quality policy is implemented and understood by all employees, and that everyone in the organisation has quality improvement objectives for their jobs. They also need to react positively to the actions resulting from quality audits.

Training at all levels within the company is required on the importance of process, product and service quality in general, and for the quality system in particular. This will help to facilitate the right type of behaviours, attitudes and values of employees towards the ISO 9000 series. Whilst providing the opportunity to answer any questions which employees may have about the QMS and the process and the reasons for registration, a systematic approach to quality, education and training will also reduce resistance to change and other obstacles.

Once all the above steps have been taken the organisation is able to commence developing its system to meet the requirements of ISO 9001. Accurate procedures, including operating and working instructions, are required. These procedures must be practical, workable and easily implemented. In writing procedures, it is worthwhile to keep in mind how to demonstrate to the auditor that the ISO 9001 requirements have been fulfilled. In simple terms this can be condensed into three principles:

- Write down what you do
- Do what you have written down
- Be able to prove it

Ownership of the procedures is important, as personnel who are given responsibilities for writing the procedures must be familiar with the requirements of the ISO 9000 series and be fully conversant with the procedures they are drafting. The use of consultants and management specialists to write procedures is undesirable as they are unlikely to fully understand the style and culture of the company. It is often found that when procedures are written in isolation and then forced upon the working environment, two main issues arise:

- Initially there is the problem of changing the way that people work without any perceived gain or benefit.
- Secondly a formal assessment of the system may reveal differences between what is written and what is done. Also, with respect to this, it is helpful to document a procedure before trying to improve it, unless the change is easy to make.

The use of others to write procedures does not allow for the positive factor of employee involvement. The ownership of the processes by those operating them is lessened. The procedures as they are being developed and/or documented need to be checked to see that they meet the requirements of ISO 9001 and how they impact other procedures, systems and activities.

The quality management system must become an integral part of the management process. When it is treated in this way, it will ensure that business improvements are incorporated into the system.

Quality Management System Assessment and Registration

When the organisation has endorsed its process controls, written the necessary procedures and instructions and developed its system to meet the requirements of the QMS for which registration is sought, the following key activities need to be accomplished.

- Train and educate staff in the workings and operation of the system and test out the procedures that have been developed.
- It may be beneficial to arrange for a pre-assessment of the system to be carried out by the selected certification body.
- Decide the most appropriate time to go for the assessment.
- ISO 9001 registration is conferred by certification bodies that have, in turn, been accredited in the UK by UKAS. The list of accredited certification bodies should be consulted, and a 'supplier audit' of them should be carried out. It is important to establish the scope of the certification body's approval, its fee structure, relevant experience and knowledge in the organisation's field of work, reputation, current workload, etc.
- The appropriate documentation is then sent to the certification body to check initial compliance against the standard. In general, a certification body will usually want to see proof that the quality system has been in effective operation for a period of six months.
- If the documentation is acceptable as it stands, some certification bodies proceed to the on-site assessment for a preliminary review (pre-audit assessment). At this stage, the company can make appropriate modifications and establish corrective actions to take into account the assessors' initial findings and comments.
- The formal assessment involves an in-depth appraisal of the organisation's quality management system for compliance with the appropriate part of the standard. This is carried out by a small team of independent assessors appointed by the certification body and generally under the supervision of a registered lead assessor, although increasingly for SMEs, only one assessor is used. At the end of the assessment, the assessors make a verbal report to management with their recommendations.
- Once the organisation is registered, the certification bodies have a system of routine surveillance. The frequency of these surveillance visits varies with the certification body but is generally twice a year. The registration usually covers a fixed period of three years, subject to the successful surveillance visits. After three years a quality system reassessment is made.

Quality Management Systems Assessment and Registration: Guidance for Small Companies

Successful registration for any organisation, especially for SMEs, requires consideration for four key factors: motivation, information, resources and planning. Organisations should assess their starting positions in each of these areas and formulate how they will make the necessary improvements to achieve successful registration.

By monitoring their progress in each of these areas, the rate of progress towards installing a quality management system to meet the requirements of ISO 9001 can be assessed.

Motivation

In a small company's journey from a primitive quality management system to attaining registration to ISO 9001, the degree of motivation can be regarded as the most important driving force. The degree of motivation can be influenced, both positively and negatively, by internal and external factors. The most powerful motivating force is demand from customers, in particular large ones, for registration and the fear of losing orders; this ensures that the momentum to introduce an ISO 9001 quality management system is maintained. Head office pressure and the impact of senior management are also factors influencing the degree of motivation, for example, in making appropriate resources available. The motivation is also affected by employee attitudes and behaviour. Antagonism or apathy from employees towards the company's efforts to obtain ISO 9001 registration makes it difficult to progress quality management system development. In addition to the primary forces there are several secondary forces, including the enhancement of company status in the marketplace, the urge to gain a commercial advantage and advertising opportunities.

Information

As well as educating the company's quality management system owner and champion on the requirements of ISO 9001, education and training on quality management system principles and practices must also be provided to inform the workforce of what is required from it. This can help to alleviate or avoid many of the problems associated with the acceptance of new working procedures, practices and disciplines. The solution for many companies is to employ management consultants to ensure that the detailed requirements of developing an ISO 9001 system are achieved, and this can help to overcome many of the problems of comprehending the requirements of the standards. By strengthening the reservoir of quality management knowledge, the quality management system champion is better able to communicate effectively with their consultants and better placed to understand the problems and pitfalls of introducing a quality management system. In this way delays in the process of documenting the system, irrelevant activities, and over-documented and bureaucratic quality manuals can be avoided. If a company is already able to build upon established quality assurance and quality control procedures (no matter how basic), this also helps to speed things up. Failure to raise the level of quality awareness and understand the demands of ISO 9001 leads to confused, frustrated and neglected employees and poorly briefed managers who will tend to restrict the progress towards registration.

Resources

There are three resources that are most significant when considering ISO 9001 certification: time, finance and availability of personnel.

In failing to make or allocate time to the process of introducing a quality management system, programmes will fall behind schedule, leading to suspension or abandonment as other, more urgent tasks appear and take precedence. Time and the availability of personnel are closely coupled. If the quality management system champion can delegate work to staff or a management consultant, time pressures can be eased. Extra resources devoted to the development of a quality management system are also instrumental in increasing the pace of progress. In small companies there is usually little slack time available in the owner's or managing director's day-to-day work activities to dedicate to the development of a quality management system. Further, there is a shortage of staff time to assist in this process, and the problem becomes more acute as the number of employees in the company diminishes. Unless compensated by a greater stimulus to produce a quality management system, a lack of time or conflicting priorities will delay or lead to the termination of progress. Also, for many small companies, the budgeting for the process of introducing and then maintaining an ISO 9001 system may require some minor restructuring of finances.

Planning

Planning is crucial to the successful introduction of an ISO 9001 system. Only by formulating a sensible plan, which details a timetable of achievable events, milestones and target dates will a company succeed in this objective. This includes recognising the need for education and training, additional skills, the use of external resources and, in some cases, the need to apprise company employees of the need for ISO 9001 registration. This domain is seen as both the magnet that draws the motivation, information and resources domains together and the glue that binds them.

Only by drawing together motivation, information and resources can progress be made towards the installation of a quality management system to meet the requirements of ISO 9001.

Benefits of Quality Management Systems

A number of benefits are associated with the implementation of a QMS. claimed for the system. As the ISO 9000 quality system series has been widely accepted throughout the world, we will once again use this QMS to illustrate the benefits that registration can bring:

- Improved controls, discipline, procedures, documentation, communication, dissemination and customer satisfaction, quicker identification and resolution of problems, greater consistency (i.e. the job is done the same way, time after time and best practices are shared), increased quality awareness, from those departments and people who traditionally perceived 'quality' not to be their major concern.
- A reduction in errors, customer complaints and non-conforming products, services and costs and the retention of customers.
- Facilitation of trade through common rules and language.
- Responsibility for quality issues is placed firmly where it belongs, with the supplier and not the customer.
- Reduction in the number of customer audits and assessments and a reduction in the time taken, leading to savings in resources needed for such activities.
- Identification of ineffective and surplus procedures and documents and other forms of waste.

- A better working environment.
- A quality management system is a fundamental pillar in an organisation's approach to QM and strategic process improvement, and it helps to ensure that any improvements made are held in place (see Figure 9.1). However, ISO 9001 registration is not a prerequisite for QM. Some organisations, particularly those from the non-manufacturing sector, have analysed and improved their systems and working practices and have then gone straight to QM.
- The guidance provided in the requirements of ISO 9001 and ISO 9004, and the independent assessment surveillance, are an indisputable aid in developing and maintaining the procedures, controls and discipline required in a quality management system. This is of value for those companies that are just commencing their quality journey.
- It is a contractual requirement of many customers that their suppliers are registered to ISO 9001; registration is also required to get on bid lists. Once a company has become registered, it is more than likely that it will ask its suppliers, distributors and providers of service to do the same, setting into motion a chain reaction. Therefore, in many sectors of industry and government procurement agencies, it is necessary from a marketing viewpoint, and without it, a company will simply not get orders. In much of the world it is now a prerequisite condition for doing business, and in some sectors of industry (e.g. the automotive industry) the rate of certification is very high. An increasing number of long-standing suppliers to companies have been told by them that they must get ISO 9001 registration to continue to be a supplier. This is despite the supplier having been the supplier of choice for a considerable period of time. Once ISO 9001 registration has been achieved, an organisation may not be able to afford to lose it.
- The preparation of systems, procedures and working instructions to meet the requirements of ISO 9001 will have a beneficial effect on a company's performance in terms of improved process yields, reduced levels of non-conformance, and improved management control.

Limitations of Quality Management Systems

On the other hand, several difficulties, problems and shortcomings have been reported and discussed. These include:

- Deciding whether registration should be sought for the whole company or just one unit/division/site/premises or even a specific operation carried out on one site or certain defined activities, as in the case of local authorities.

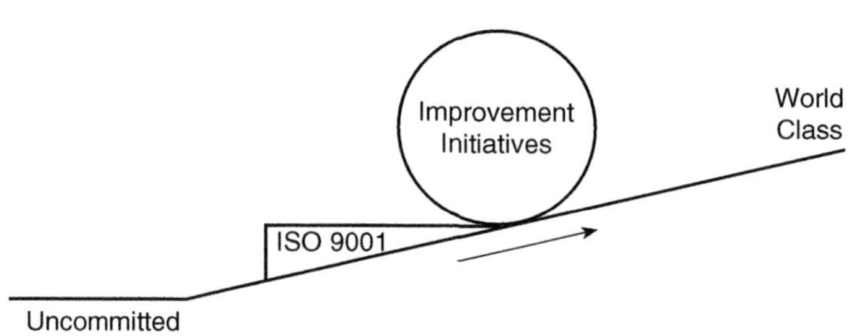

Figure 9.1 Quality improvement and the ISO 9000 series.

- Applicability of the standards to certain situations, particular sectors of business, and management styles.
- Interpretations of various sections of the standard and understanding the requirements of the standards.
- Terminology used.
- Lack of flexibility and perceived restrictions on creativity.
- Lack of relevance to the real needs of the business, resulting in a view that it was 'bureaucracy gone mad' and a 'why bother' attitude from operational personnel.
- The time and resources needed for writing procedures and training and retraining staff in the requirements of the ISO 9000 series and the internal auditing of the system.
- The bureaucracy involved in documentation and accreditation, and the lack of mutual recognition of certificated bodies between countries.
- The cost involved in achieving ISO 9001 registration and then maintaining it. This applies to small companies. The cost comprises the additional workload incurred by company personnel in writing the procedures, managerial time, increased paperwork, etc., the fee of the management consultancy (if a consultant is used to assist with the process of registration) and the certification body's fees.
- The perception of some SMEs that this is only applicable to large companies.
- Considered by those companies that have mature QM approaches to be of no value.
- In some cases, in particular sales/service situations, the rigour and applicability of the standards are perceived as restrictive and as barriers to providing a flexible and responsive service to customers.
- Lack of internal and external audit rigour.
- It is not easy to get every function and person involved to take responsibility for their own quality assurance and to make quality improvements in the processes for which they are responsible.
- Suppliers have a habit of doing what their customers want, and many organisations have achieved ISO 9001 registration to provide documented proof that they have an adequate quality system in place just to satisfy the demands of their major customers. This may not produce the required improvement culture naturally, and any gains made will be short-lived if registration is perceived as a contractual condition rather than a foundation for ongoing improvement.
- A system based on the ISO 9000 series provides only the foundation blocks, and registration to ISO 9001 should be viewed as the minimum requirement; the objective should be to develop and improve the system in relation to the needs of the organisation. An organisation does not achieve superior-performing company status merely by ISO 9001 registration. It is clearly a pre-competitive issue, and separate from the ability to compete, which depends on many other factors. The winners will be those who have a dedicated commitment to company-wide improvement through continuous self-assessment of what they do.
- The mechanisms of the ISO 9000 series are such that they will tend towards a steady-state performance. The ISO 9000 series of standards is designed to produce consistency in actions, products and services. An organisation can have consistent performance with a high level of non-conformance.
- Experience indicates that the ISO 9000 series can have a limited impact on the overall improvement of an organisation simply because it does not get to the root cause of problems. Most problems are resolved at the branch level, and this is a failure in several businesses.

- In some quarters there is confusion about the relationship between the ISO 9000 series and QM or strategic process improvement. They are not alternatives; a quality system is an essential feature of QM. However, some organisations see ISO 9001 registration as the pinnacle of their QM achievements, and no plans are laid for building on this registration; a small number of people even believe that improvements driven through internal audits of the ISO 9001 will lead their organisation to QM. As previously mentioned, registration often results in a sense of complacency after a successful third-party assessment of the system.
- ISO 9001 registration or any other QMS registration will not prevent a supplier from producing and delivering non-conforming products and/or services to its customers. The standards are a specification for the management of quality; there is a clear distinction between registration and capability, and this fundamental fact needs to be recognised. Product and/or service quality is determined by the individual organisation and its people and processes, and not by a quality management system standard.

Summary

A quality management system is one of the key building blocks for an organisation's QM and strategic process improvement activities. ISO 9001 and ISO 9004 define and set out a definitive list of features and characteristics which should be present in an organisation's quality management system through documented policies, manuals and procedures, whatever the product or the service, and provide a framework for how an organisation may develop a quality system. The design of a QMS must maximise ownership, allow flexibility without loss of control, and be able to be developed to cope with changes in the business and capture improvements; above all it must be 'user-friendly'.

Seeking registration for the wrong reasons and a system that is too inflexible and bureaucratic are some of the major pitfalls. Assessment to ISO 9001 may improve an organisation's systems, procedures and processes, but on its own will not deliver continuous and company-wide improvement. To make the best use of the ISO 9000 series it is important that the implementation is carried out in the right spirit and for the right reasons. This is an area in which management commitment is vital. Many of the reported difficulties, shortcomings and criticisms lie in the hands of an organisation's senior management team. The saying 'you only get out what you put in' is so relevant to the development of a QMS, and it is so important that the system is seen as being alive. All too often the QMS is left solely in the hands of the quality department, often just one individual.

Registration to ISO 9001 is not the only way to achieve quality assurance, nor is it a prerequisite for QM or strategic process improvement. It is, however, sometimes necessary to have the appropriate registration to do business at both a national and an international level, and in this respect, it is a key marketing tool. It is the fear of loss of business and substitution in the marketplace that has caused many organisations to obtain ISO 9001 registration. The ISO 9000 series provides a common benchmark for good-quality management system practice, which is recognised throughout the world. An organisation that is registered to ISO 9001:2015 should be working in an organised, structured and procedural way with defined methods of operating. It is important that organisations do not view ISO 9001 registration as their pinnacle of success in relation to quality assurance and quality management. It only provides the foundation blocks, and they must have strategies and business plans in place to move on and cater for areas which are not addressed by the standard and develop to QM. This is particularly important in smaller businesses, which, in a number of cases, attain ISO 9001 registration and have no interest in or vision of developing further their quality management activities.

References

Bamford, D., Forrester, P., & Reid, I. (2023). *Essential guide to operations management: Concepts & case notes*, 2nd edition. UK: Routledge.

ISO (2023). *The ISO Survey of Management System standard certifications*. Geneva: International Organisation for Standardization.

ISO 10013 (2021). Quality management systems – Guidance for documented information. Geneva: International: Organisation for Standardization. https://www.iso.org/standard/75736.html

ISO 14001 (2015). *Environmental management systems: Requirements with guidance for use*. Geneva: International Organisation for Standardization.

ISO 19011 (2018). *Guidelines for auditing management systems*. Geneva: International Organisation for Standardization.

ISO 45001 (2018) *Health and Safety Management Systems*. UK: Health and Safety Executive, . https://www.hse.gov.uk/managing/management-system/iso45001.htm

ISO 9000 (2015). *Quality management systems: Fundamentals and vocabulary*. London: British Standards Institution.

ISO 9001 (2015). *Quality management systems: Requirements*. London: British Standards Institution.

ISO 9004 (2018). *Managing for the sustained success of an organisation: A quality management approach*. London: British Standards Institution.

Sherwani, K.H., Demir, A., & Maroof, L. (2025). Way to achieve sustainable benefits of ISO 9001 practices. *International Journal of Quality & Reliability Management*, 42(4), 1167–1195.

Warner, F. (1977). *Standards and specifications in the engineering industries*. London: National Economic Development Office.

Part 4

Quality Management Through Engagement

10 Teams and Teamwork

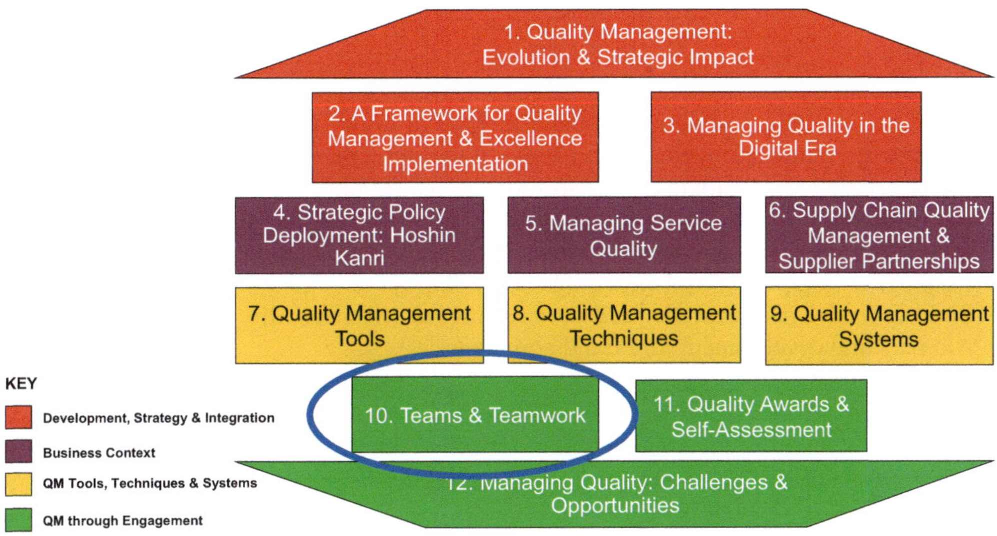

Introduction

The development of employees and their active participation in improvement activities, both individually and collaboratively, is a key feature of a company's approach to Quality Management and Strategic Process Improvement. A fundamental aspect of this approach is the role of management in effectively utilising employees' skills and expertise while simultaneously developing the problem-solving and continuous improvement capability of the company. This process not only enhances individual growth but also contributes to the overall results of the organisation and the creation of a continuous improvement culture.

There are several different types of teams with different operating characteristics, all of which can act as a vehicle for getting people involved in improvement activities and improving organisational performance. Teams can be found everywhere and for almost everything, and most organisations have them. Some teams have a narrow focus, with members coming from one functional area; others are wider and cross-functional, dealing with the deep-rooted problems between internal customers and suppliers. Each type of team has its advantages.

In some instances, management may initiate team-based improvement activities by assembling a group of employees and expecting them to function effectively without

DOI: 10.4324/9781003594369-14

providing adequate training, coaching, guidance, managerial support, or team-building efforts. Under such conditions, it is unsurprising that the team may struggle to maintain effectiveness and deteriorate within a few months of its formation.

This chapter, which is based on research on teams and practical experience of working with teams in many different types of organisations, examines the role of teamwork and outlines the operating characteristics of project teams, quality circles, and quality improvement teams. The key constituents of teams in terms of sponsor, facilitator, leader, and member are outlined. Some guidelines are also given, which should help to ensure that teams are both active and effective.

The Role of Teams in Continuous Improvement

Teams have a few roles to play as a component in a process of continuous improvement. Teams can:

- Aid the commitment of people to the principles of Quality Management.
- Improve communication within and across functions, and with customers and suppliers.
- Allow people to participate in decision-making about how the business operates.
- Improve morale, relationships, and knowledge, develop trust, facilitate cooperative activity, problem-solving, and adjustment to change.
- Help to develop people and encourage leadership traits.
- Build collective responsibility and develop a sense of ownership.
- Aid personal development and build confidence.
- Facilitate awareness of quality improvement potential, leading to behaviour and attitude change.
- Improve the integration of new products into existing production and service processes.
- Improve operational efficiency by aligning team efforts towards shared objectives, thereby developing interaction and synergy.

Types of Teams

High-performing organisations maximise the benefits of teamwork, with senior management and cross-functional teams working cooperatively on Quality Management activities such as benchmarking, supplier development, etc. This collaborative approach enhances efficiency, innovation, and continuous improvement across the organisation. Some teams may be hierarchical in nature – a corrective action team is formed on a directive from a quality improvement team. For teams to operate effectively, top management must demonstrate the expected team behaviours and collaborative work practices, setting a clear example for the organisation. High-performing organisations employ various methods to strengthen team dynamics, enhance workplace relationships, and reinforce a culture of teamwork.

Having made the case for teams, it is important for the management of an organisation to decide when and how to use teams, and what the appropriate conditions are that need to be present. There are many types of teams with differing characteristics in terms of membership, autonomy, scope of activity, decision-making authority, access to information, problem-solving potential, resources, and life span, which can be used in the improvement process. It is important that the right type of team is formed for the project, problem or activity under consideration and that a working definition of the team is decided upon. Some of the most popular types of teams are discussed below.

Project Teams

As discussed in Chapter 1, the drive to improve quality originates at the top of an organisation. Once the main problems facing the organisation are identified, they can be developed into a portfolio of one-off projects. The project owner will select the team members and can choose to either lead the team themselves or act as a 'sponsor' to the team. Through participation in project teams, managers better understand the problem-solving process and become more sensitive to the problems faced by other types of teams. The typical characteristics of such teams are:

- The objective has been defined by senior management.
- The team is led by management.
- It is temporary in nature.
- The project is specific and significant, perhaps addressing issues of strategic change, and will have clear deliverables within a set timescale.
- The team is organised to ensure that it employs the appropriate talents, skills, and functions that are suitable for the resolution of the project.
- The scope of activity tends to be cross-functional.
- Participation is not usually voluntary – a person is requested by senior management to join the team, and this is done based on their expertise in areas related to the project being tackled.
- Team meetings tend to be of long rather than of short duration, although they occur on a regular basis.

Quality Circles

Quality Circles (QCs) have characteristics that are different from other methods of teamwork. They have been the subject of much research (e.g. Li and Doolen, 2013; Seo et al., 2015; De Menezes & Escrig, 2019). The QC approach was established by Japanese manufacturers in the aftermath of World War II in response to the poor quality of Japanese products in international markets (Ellis et al., 2019). The implementation of QC was regarded as being instrumental in improving the quality of Japanese products in international markets, contributing to Japan's rapid rise as a major global economy. (Ishikawa, 1985).

Western interest in Japanese quality methods gained popularity in the 1970s, and QCs were introduced in Western companies in the early 1980s with the aims of improving employee participation in continuous improvement, improving Quality Management, and changing company culture. By the end of the 1980s the QC approach in Western companies waned, leading some to dismiss it as a fad. Challenges included technical issues, managerial resistance, and ineffective structures. (Dale et al., 2016). However, QCs in many Japanese companies have continued to play an important role in Quality Management.

Toyota and Quality Control Circles

'The primary goal of Quality Control Circles (QCCs) is to Motivate and Engage the workplace through member development and engagement in problem solving activity' – according to Kevin Robinson, Former Assistant General Manager of Toyota (2022). The characteristics of QCCs at Toyota are:

- Participation in QCC activities is expected at all levels of the organisation, from senior management to frontline team members.
- Each QCC cycle spans approximately six months from initiation to completion.
- QCC is made up of five team members, usually from the same section or area of production.
- Team Leaders take the role of coach and facilitator of QCC and are required to complete two QCC cycles per year, while individual team members are encouraged to participate in at least one QCC biannually.
- The Production Plan includes a scheduled 30-minute planned stop every two weeks to facilitate QCC activities. This designated time also allows managers who are experienced in problem-solving to actively support QCC teams.
- QCC teams are permitted to allocate additional time to further develop their problem-solving efforts beyond the scheduled sessions.
- QCC follows an A3 Problem-Solving template, which incorporates an 8-step process methodology.
- Teams independently select their problem statements, typically beginning with simpler challenges in their initial QCC cycle.
- As employees gain experience with QCCs, their problem-solving capabilities improve, enabling them to address more complex challenges that align with the organisation's strategic objectives.
- QCC initiatives address various areas, including safety, quality, productivity, and environmental sustainability, all of which contribute to the long-term competitiveness of the organisation.
- At the conclusion of each QCC cycle, teams present their findings. The most outstanding QCC team within the plant advances to a regional competition, where top-performing teams compete for the opportunity to present their solutions in Japan, sharing best practices in the organisation.

Team Improvement Circles

Team Improvement Circles (TICs) represent a structured, team-based initiative like QCs teams, which have gained prominence in the 2010s. An example of this approach is Jaguar Land Rover, a UK-based automotive manufacturer specialising in premium vehicles, which introduced TICs as part of its commitment to continuous improvement. Sponsored by senior management, the primary objective of TICs is to increase employee engagement and development through structured problem-solving methodologies.

Typically, TICs consist of three to five volunteer members and are encouraged to be cross-functional, incorporating participants from wider business functions to increase collaboration and innovation. The scope of TIC projects is determined based on key criteria, including customer impact, integrity, personal growth, feasibility, and overall impact. To ensure systematic problem-solving and sustained improvements, TICs adhere to a structured eight-step process that guides their initiatives from identification to implementation.

- Step 1 Form a Team
- Step 2 Look for Improvement Themes

- Step 3 Select A Theme
- Step 4 Baseline Conditions
- Step 5 Set Vision
- Step 6 Gap Analysis
- Step 7 Create an Action Plan
- Step 8 Sustain, Improve, and Results
- Step 9 Reassessment of Circle
- Step 10 Present

Starting in 2015 with 75 completed TICs, this had become a global initiative with over 2,000 TICs delivered in 2024. The best TICs are judged in a regional competition, then a global competition.

Quality Improvement Teams

These teams may be composed of members from a single department, operate as cross-functional units, and include representatives from customers, suppliers, or both. Their primary objectives are typically centred on addressing quality challenges or implementing improvements in response to customer and stakeholder requirements. This may involve enhancements in safety, quality, cost efficiency, delivery performance, or environmental sustainability.

The characteristics of quality improvement teams are more varied than for any other type of team, but typically include:

- Process Owners – Individuals responsible for overseeing specific business processes and ensuring improvements align with organisational objectives.
- Subject Matter Experts (SMEs) – Employees with deep knowledge in specific areas who provide technical insights and recommendations.
- Team Leaders or Facilitators – Individuals trained in continuous improvement methodologies who guide discussions, facilitate problem-solving, and ensure progress.
- Cross-Functional Representatives – Members from different departments (e.g., operations, quality control, supply chain, and customer service) who contribute insights from their respective areas.
- Senior Management Sponsors – Leaders who support initiatives, allocate resources, remove bottlenecks, and ensure alignment with organisational priorities.

These teams can be voluntary or mandatory, and work on specific projects as a result of a management initiative, a need to undertake some form of corrective action, a high incidence of defects, supplier/customer problems, or opportunities for improvement.

Continuous Improvement Teams

Continuous Improvement (CI) Teams are specialised groups within an organisation dedicated to enhancing overall performance through the application, promotion, and support of Quality Management, Lean, and Six Sigma principles. CI team members are often integrated into other project-based initiatives, such as project teams and quality teams, where their knowledge of continuous improvement tools can be effectively applied. The primary function of these teams is to drive systematic enhancements in operational processes, product quality, and service delivery. These teams focus on process improvement, increasing efficiency, and eliminating non-value-added activities. Their initiatives are typically aligned with broader organisational objectives, including cost reduction, customer satisfaction, and operational excellence.

High-performing teams progress through distinct stages of development and require an effective balance of roles and team dynamics to achieve success. Tuckman's (1965) model of group development, later expanded by Tuckman and Jensen (1977), identifies four key stages: forming, storming, norming, and performing. This framework remains crucial in understanding team dynamics within Quality Management. Additionally, a diverse combination of team roles is essential for effective collaboration. Belbin's (2010) research on successful and unsuccessful teams highlights the necessity of varied roles, identifying nine key types: Plant, Monitor Evaluator, Coordinator, Resource Investigator, Implementer, Completer Finisher, Team Worker, Shaper, and Specialist (see www.belbin.com for more information). The following team roles are suggested.

Team Sponsor

- Quality steering group (QSG) member and a senior manager.
- Actively supports the team in its task, especially by contributing to the removal of roadblocks, and helps to resolve priority conflicts.
- Mentor to the team leader.
- Ensures that the team leader has the skills and training required to lead the team.
- Communications link between the team and QSG and with departments that are affected by the project.
- Ensures that the team's activities are accepted by the department in which it is working (e.g. holds meetings with the area management, staff and operator to help to generate a total understanding of the project and the reasons for it).
- The sponsor agrees a charter (Objectives, outcomes, resources, boundaries, time scales) with the team.
- Ensures that other people and teams are not addressing the same project as this team.
- Holds regular meetings with the team's facilitator.

Team Facilitator

- Helps the team mentor and team leader establish the team.
- Ensures the right balance of skills is present within the team.
- Acts as coach to the team leader.
- Responsible for team progress and direction.
- Ensures all team members contribute.
- Responsible for communication from the team to the outside world.
- Communicates team roadblocks to the team sponsor and helps remove them.
- Identifies training needs of the team and provides and implements solutions, as appropriate.
- Assists the team in preparing recommendations and presentations to management.
- Helps sponsor resolve external resource/priority conflicts.
- Celebrates success with the team.

Team Leader

- The leader of a team can be chosen by management or appointed by the team. Another issue to be considered is whether the role of team leader should be rotated amongst team members.
- Organises and sends out an agenda for meetings.
- Ensures the team members are familiar with the protocol of team meetings.
- Needs to be an active contributor and listener.
- Ensures that team members know what is expected of them.
- Leads the definition and implementation of team processes.
- Responsible for team progress and direction.
- Identifies any training needs of the team and its members.
- Provides regular verbal reports and copies of team minutes to the mentor about the progress of the team.

Team Member

- Needs to be enthusiastic about the project; its resolution must be of direct benefit to them.
- Must be a willing team member; not coerced into joining the team.
- Should contribute relevant experience.
- Is prepared to commit time outside team meetings to collect data and carry out agreed actions.
- Takes responsibility, as requested, for follow-up actions.
- Respects the role of team leader.
- Needs to be an active contributor and listener.
- Needs to be able to follow through actions.
- Should respect the ideas and views of other members.

Evaluation of Teams

It is not easy to evaluate the effectiveness of teamwork, other than by the effectiveness of the actual solutions produced and improvements made. It is, however, important that the 'health' of a team is regularly assessed. The observable characteristics of an effective and ineffective team are (Table 10.1).

Team Competition

Like the previously discussed QCs and TICs, it is the formal recognition and celebration of team achievements, as well as the promotion of role model behaviours. A commonly employed approach is the organisation of an annual team competition or conference, typically conducted off-site, where exemplary team initiatives are showcased. A judging panel, comprising both internal personnel and external experts, evaluates the projects and their presentation, based on key criteria, including theme selection, problem analysis and resolution, team participation and contributions, and achieved outcomes and benefits (Box 10.1).

Table 10.1 The observable characteristics of an effective and ineffective team

Observable characteristics of an effective team	*Observable characteristics of an ineffective team*
• Everyone is participating, making a contribution and involved in actions and through this is achieving their personal potential. • Relationships are open. • Team members trust, respect and support each other and are prepared to adapt and be cooperative. • Members listen closely to the views of other members of the team, have an open mind, and maintain a positive attitude. • Everyone expresses their views, ideas, and problems, and all available means are used to support ideas. • Members respect the operating procedures and principles of the team, and they own the team process. • There is clarity of focus on the project being tackled, and members know what is expected of them. • The team leader has the ability to translate ideas into action.	• Poor leadership. • Cliques, defensiveness, closed minds and blame culture within the team membership. • Downright hostilities, conflict, competition and lack of tolerance between team members. • Members are not all participating in the activities of the team. • Limited communications between team members, and members have a tendency to act on their own. • Insufficient attention to the team process. • There is no pride displayed in the team activity. • Members feel they are being taken advantage of and the higher performers reduce their efforts to those of the lowest performer.

Box 10.1 Annual Team Competition (Adapted from Dale et al., 2016)

For example, an organisation holds its own internal event to decide which quality improvement team will represent it at the annual national competition. Each site team submits a project brief listing team members, project objectives and details, how data is analysed, the problem-solving approach used, results and outcomes, and future opportunities.

Each member of a team that makes it through to the final receives a certificate and a small financial reward to be spent on a team development activity such as a dinner, attendance at a sports event, theatre visit, attendance at a training event, etc. The winning team gets a similar certificate and reward, but in addition, the annual trophy.

This competition had real impacts on business performance - failure rates halved, and productivity increased by 35–40%.

Guidelines for Developing Effective Teams

As outlined in Chapter 1, a CI process will encounter periods of stagnation when nothing appears to be moving. This phenomenon is mirrored by teams whose members begin with high levels of energy but can quickly slump and suffer frequent periods of inactivity.

To ensure that teams work effectively and efficiently, the following factors should be considered:

- Management must commit themselves to nurturing and supporting teams. They need to release, on a gradual basis, authority, decision-making and accountability and put in place a suitable organisational support structure and connections between the team and the individuals and functions which it affects.
- Prior to launching any form of team activity in relation to the introduction and development of Quality Management, it must be ensured that the appropriate awareness training and education about Quality Management have been undertaken.
- Each member must be clear about the aims and objectives of the team, the work to be accomplished by the team, and its potential contribution to day-to-day operations.
- The team should have specific goals and an action plan with completion dates related to the objective. This not only assists in setting boundaries for the project but also helps the team to stay focused on it. The project must not be too large to discourage team members; an early success is vital.
- Team members must be trained in appropriate and robust data collection, problem-solving, and decision-making methods. Training should include project planning, team building, and team dynamics to provide an understanding of the behavioural needs which may determine team effectiveness.
- Special coaching and counselling should be provided to the team leaders since they are critical to a team's success. They must have the appropriate leadership skills in relation to the team.
- The team mentors must be seen to be actively supporting and contributing to the team. This applies to requests for resources and support from key organisational functions.
- Teams should meet on a regular basis and work to an agenda. The team must never leave a meeting without agreeing on future actions and the date and time of the next meeting. Once a team meeting has been agreed, only in exceptional circumstances should the leader or any team member fail to attend the meeting.
- Regular reports to management and the 'mentor' on team activities must be prepared. The results and decisions should be communicated accurately to the rest of the workforce. It is also important that management carry out a periodic review of each team's progress.
- People who are likely to be affected by the results of the project should be involved in the team activity.
- The team should receive appropriate recognition and celebration for successful improvements.
- The performance of the team on completion of a project should be evaluated and reviewed to see what worked, what did not, and what could be done better, including the identification of training needs and pinpointing of barriers.

Summary

Teamwork is a key element of any Quality Management and Strategic Process Improvement approach. There are a variety of teams with different operating characteristics that can be used in Quality Management and Strategic Process Improvement. Superior-performing organisations employ several types of teams. Different types of teams can be used at different stages of an organisation's development of Quality Management. Some teams are drawn from one functional area of the business and have a narrow focus with perhaps limited problem-solving potential. Other types of teams are wider in focus and tend to be cross-functional. The type of team will depend on the nature of the problem and objectives, and management must decide which type is needed. This chapter has concentrated on four types – project teams, quality circles, quality improvement and CI teams – and described their operating characteristics.

It is surprising how many organisations make several fundamental mistakes in establishing teamwork as part of their Quality Management and Strategic Process Improvement approach:

- Teams are not given any training prior or during the activity.
- The wrong type of team is established for the project being tackled.
- The team is structured in such a way that members discuss their views on the cause of the problem, and these ideas are then passed over to technical personnel and engineers for them to come up with a solution: the result is that team members feel they have achieved nothing and become disaffected with the team process.
- Too many teams are introduced at one time, which the infrastructure cannot support.
- The leader is unaware of the importance of their role to the success of the team.

To help organisations avoid some of the common mistakes, guidelines have been outlined that should be considered prior to setting up any form of team activity. However, even if these are followed, teams are likely to encounter periods of stagnation when nothing appears to be happening. The importance of assessing the health of teams has been discussed to help team members analyse strengths and areas for improvement and thereby overcome these periods of inertia and maintain performance and effectiveness. It is vital that teams learn from experience and make the necessary changes for the good of the team. It must be remembered that it takes hard work and commitment to develop effective teamwork.

Teams and groups are important, as has been outlined in this chapter, but their effects must not be overemphasised in relation to individuals working alongside on organisational improvement initiatives.

References

Belbin, R.M. (2010). *Management teams: Why they succeed or fail*. London: Heinemann.

Dale, B., Bamford, D., & van der Wiele, T. (2016). *Managing Quality: an essential guide and resource gateway*, 6th edition (p. 330). UK: John Wiley & Sons.

De Menezes, L.M., & Escrig, A.B. (2019). Managing performance in quality management: A two-level study of employee perceptions and workplace performance. *International Journal of Operations & Production Management*, 39(11), 1226–1259.

Ellis, G., Found, P., Kumar, M., & Harwell, J. (2019). New evidence on the origins of quality circles. *Total Quality Management & Business Excellence*, 30(sup 1), S129–S140.

Ishikawa, A. (1985). Principles of QC circle activities and their effects on productivity in Japan: A corporate analysis. *Management International Review*, 25(3), 33–40.

Li, J., & Doolen, L.T. (2013). A study of Chinese quality circle effectiveness. *International Journal of Quality & Reliability Management*, 31(1), 14–31.

Robinson, K. (2022). Talk on 'Lean and Operational Excellence', at 'Journey to Excellence' conference in Manchester, 2 September, 2022.

Seo, Y., Lee, C., & Moon, H. (2015). An organizational learning perspective of knowledge creation and the activities of the quality circle. *Total Quality Management & Business Excellence*, 27(3-4), 432–446.

Tuckman, B.W. (1965). Development sequence in small groups. *Psychological Bulletin*, 63(6), 384–399.

Tuckman B.W., & Jensen, M.A.C. (1977). Stages of small group development revisited. *Group and Organization Studies*, 2(4), 419–427.

11 Quality Awards and Self-Assessment

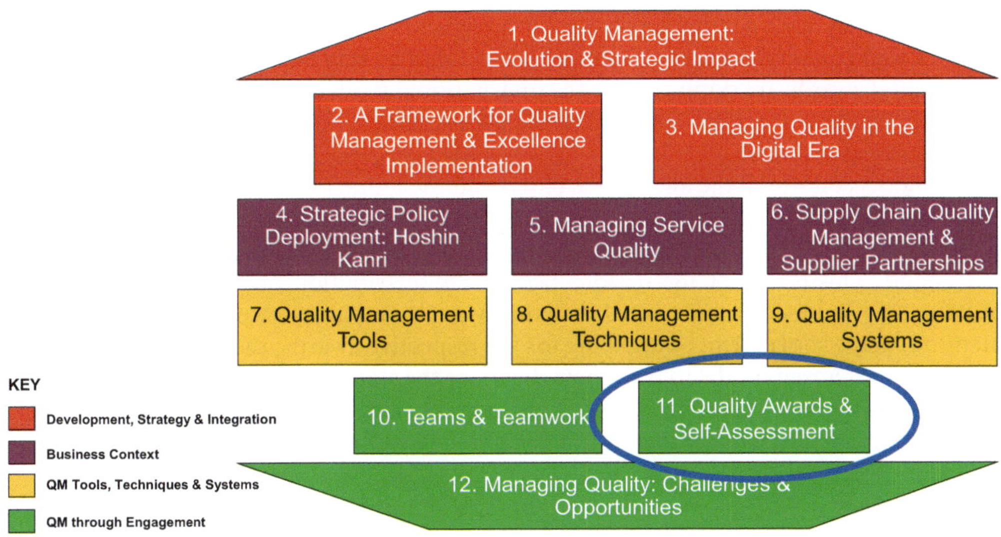

Introduction

To sustain and continually improve a quality and excellence culture, organisations must regularly monitor their performance, identify gaps, and take action to address them. Self-assessments provide a framework for generating this type of feedback about an organisation's approach to Quality Management (QM).

Self-assessment against the criteria of a quality award/excellence model on which evaluation and diagnostics can be based is now being given a considerable amount of attention by organisations throughout the world. The criteria of these awards encapsulate a comprehensive and holistic management model covering its various activities, practices, and processes, and provide the mechanism for quantifying an organisation's current state of QM development by means of a points score. There are many definitions of self-assessment provided by Dale et al. (2016); however, one that demonstrates both the advancement of QM and continuous improvement across sectors is from the Police Service of England and Wales. The services utilise self-assessment as their method for empowering leaders in

DOI: 10.4324/9781003594369-15

service-led improvement and cultural change. They (College of Policing, 2025) define self-assessment as:

> [to] clearly show what is accepted by practitioners as good practice to achieve a sustainable continuous improvement culture; help forces identify their own areas of strength, as well as areas for improvement; allow forces to benchmark their current continuous improvement capability and plan which areas they will develop as part of their business change planning.

There are several internationally recognised models, the main ones being the Deming Prize in Japan, the Malcolm Baldridge National Quality Award (MBNQA) in America, the EFQM Global Award, and the Shingo Prize. Although there are some differences between the models, they have a number of common elements and themes. In addition, there are many national quality/excellence awards and regional quality awards. Most of the national and regional awards are duplicates of the international models, with some modifications to suit issues that are of national or local interest. In America alone there are over 40 state and regional award schemes.

The models on which the awards are based and the guidelines for application are helpful in defining QM and Organisational Excellence in a way that the management of all types of organisations – small, large, public, private, manufacturing, and service – can easily understand. This is one of the reasons behind the distribution of thousands of booklets outlining the guidelines and award criteria. Many companies requesting them have no intention, in the short term, of applying for the respective awards; they are simply using the criteria of the chosen model to help them diagnose the state of health of their improvement culture and provide indications of how to achieve business excellence. They help organisations to develop and manage their improvement activities in several ways. For example:

- They provide a definition and description of QM and Organisational Excellence, within a defined framework, which gives a better understanding of the concept, improves awareness, and generates ownership amongst senior managers.
- They enable measurement of the progress with QM and Organisational Excellence to be made in a structured and systematic manner, along with its benefits and outcomes.
- Regular improvement is encouraged, and this provides the basis for assessing the rate of improvement.
- They force management to think about the basic elements of their business and how it operates, the relationship between actions and results, and how organisational change is facilitated: 'Where are we now and where do we need to be in the future?'
- The scoring criteria provide an objective, fact-based measurement system, and help gain consensus within the organisation on the strengths and areas for improvement of the current approach and pinpoint the key improvement opportunities.
- Sharing of best practices and organisational learning is facilitated.
- Education of management and employees on the basic principles of QM and Organisational Excellence is improved.
- They help develop a more cohesive company working environment.

Based on detailed research Dale et al. (2016) have summarised the benefits of self-assessment against a quality/excellence award model (see Table 11.1).

Table 11.1 Benefits of the self-assessment process

Category	*Benefits*
Immediate	• Facilitates benchmarking • Drives continuous improvement • Encourages employee involvement and ownership • Provides visibility in direction • Raises understanding and awareness of quality-related issues • Develops a common approach to improvement across the company • Used as a marketing strategy, raising the profile of the organisation • Produces 'people-friendly' business plans
Long-term	• Keeps costs down • Improves business results • Balances long- and short-term investments • Provides a disciplined approach to business planning • Develops a holistic approach to quality
Supporting QM	• Increases the ability to meet and exceed customers' expectations • Provides a link between customers and suppliers Supporting Quality Management • Helps to refocus employees' attention on quality • Provides a 'health check' of processes and operations • Encourages a focus on processes and not just the end-product • Encourages improvements in performance

To use any self-assessment method effectively as a business tool for continuous improvement, various elements and practices must be in place, and management needs to have had some experience with QM or Organisational Excellence to understand the questions underpinning the concept. What has not been implemented cannot be assessed. The decision to undertake self-assessment needs to be fully considered from all angles, and management must be fully committed to its use. It is the authors' view that the use of self-assessment based on a quality or excellence model depends on the organisation's current level of QM and improvement maturity. Many awards are most appropriate for organisations that have spent several years implementing QM and continuous improvement methodologies before evaluating themselves against a recognised quality or excellence framework. However, the EFQM positions its model as one that supports new organisations by providing a roadmap to help management develop a stakeholder-driven strategy focused on creating sustainable value.

Having made this point, the models underpinning the quality/excellence awards are also helpful in demonstrating what is involved for managers in organisations that are not experienced in QM. However, they must understand the potential gap that can exist between where they currently stand in relation to QM and the model of the award being used in order to make comparisons.

This chapter opens by investigating the replacement of Strategic Process Improvement and QM by excellence in the award models, and this is followed by a description of the Deming, MBNQA, EFQM, and Shingo Prize models. It then examines the self-assessment process and the associated success factors and difficulties.

Quality, QM, Strategic Process Improvement, and Excellence

Over the years there has been a drive to change from 'Strategic Process Improvement' and 'Total Quality Management' (TQM) to 'Excellence' in the criteria of the quality/excellence models. This originated with the EFQM and management consultancies, the former in response to the perceived tarnished image of TQM and Strategic Process Improvement initiatives, and the latter seeing a diminishing demand and/or increasing competition for their services. In response to the 'fallen star' image of TQM and Strategic Process Improvement initiatives, the EFQM, in their excellence model, has progressively stripped out reference to TQM and quality in both the criteria and the areas to address.

Fundamental concepts of excellence are (EFQM, 2025):

- Adding value for customers
- Creating a sustainable future
- Developing organisational capability
- Harnessing creativity and innovation
- Leading with vision, inspiration, and integrity
- Managing with agility
- Succeeding through the talent of people
- Sustaining outstanding results

In the 2015 revision of the ISO 9000 series, the principles of QM are defined in BS EN ISO 9001 (2015) as:

> Customer focus; leadership; engagement of people; process approach; improvement; evidenced based decision making; relationship management.

Allowing for interpretation of these individual principles, there is little or no difference between QM and excellence. This lack of clear water in definitions is confirmed when comparison is made with the 10- and 14-point cluster summaries of the teachings of Crosby, Deming, Feigenbaum, and Juran, and with what has been written about earlier in this book as the elements and practices of TQM.

The Award Models

Deming Prize

Established in 1951 by the Union of Japanese Scientists and Engineers (JUSE), the Deming Prize is one of the oldest awards in QM. Named in honour of American quality expert Dr. W. E. Deming, it recognises his contributions and lasting friendship with post-war Japan in advancing QM practices in the country (Deming Prize Committee, 2000).

Unlike many other awards, the Deming Prize does not require applicants to follow a specific model, system, method, or technique. Instead, it adopts a holistic approach to deep learning and the implementation of TQM. Applicants are expected to assess their own situations, define relevant themes and objectives, and drive organisation-wide improvement and transformation. This approach is intended to develop a deep, shared understanding of TQM across all employees.

Evaluation focuses on whether the applicants' chosen themes align with their specific circumstances, whether their initiatives are appropriate to those circumstances, and whether these efforts are likely to achieve higher objectives over time. In Japan, the award supports the discussion on how to apply the TQM approach more vigorously through a common core level of understanding of TQM. According to JUSE (2025), the benefits of applying for the Deming Prize include:

- New product development and launch will be promoted
- New technology can be obtained
- Sales target will be achieved continuously
- Become capable of following a policy and fulfilling an objective
- Organisation's functionality will be enhanced

There are five separate divisions for the award: the Deming Prize, the Deming Prize for Individuals, the Deming Distinguished Service Award for Dissemination and Promotion (Overseas), the Deming Distinguished Practice Award, and the Deming Grand Prize. The Deming Prize is open to individual sites, a division of a company, small companies, and overseas companies.

The Deming Prize was originally limited to Japanese companies but has been open to international companies since 1984. From 2015 to 2024, approximately two-thirds of organisations receiving the prize were based outside Japan. India accounted for 40% of entries during this period, followed by Thailand, China, Indonesia, and the USA.

The Deming Prize is an annual award presented to an organisation that has implemented TQM suitable for its management philosophy, scope/type/scale of business, and management environment. The Deming Grand Prize is an annual award presented to an organisation that received the Deming Prize three or more years ago, that has continuously applied TQM to priority issues and has been achieving its objectives steadily and effectively in a changing business environment (JUSE, 2025).

The Deming Prize consists of three primary categories (see Table 11.2), which are divided into a total of 11 sub-categories. JUSE selects the examiners for the Deming Prize from QM experts from not-for-profit organisations. The applicants are required to submit a detailed document on each of the prize's criteria. The size of the report is dependent upon the number of employees in each of the applicant company's business units, including the head office. The Deming Prize Committee examines the application document and decides if the applicant is eligible for on-site examination. The committee chooses two or more examiners to conduct this examination. Discussions with JUSE suggest that considerable emphasis is placed on the on-site examination of the applicant organisation's practices. It is also evident that the applicant organisation relies a great deal on advice from the JUSE consultants. JUSE would also advise an organisation when it should apply for the prize.

The Malcolm Baldrige National Quality Award

The MBNQA recognises US organisations in the business, health care, education, and nonprofit sectors that excel in QM and performance excellence. It was created in a bid to improve the QM practices and competitiveness of US firms. The Malcolm Baldrige National Quality Improvement Act of 1987 established this annual US quality award, some 37 years after the introduction of the Deming Prize. The award is named after a former American Secretary of Commerce, Malcolm Baldrige, who served from 1981 until his death in 1987. The purposes of the award are to raise awareness of quality and performance management and to recognise US organisations that have successfully implemented

Table 11.2 Quality award criteria

Category	*Max*
Deming Prize	
Management policies and their deployment regarding quality management	20
New product development and/or work process innovation	20
Maintenance and improvement of product and operational qualities	20
Establishment of systems for managing quality, quantity, delivery, costs, safety, environment, etc.	10
Collection and analysis of quality information and utilisation of IT	15
Human Resources development	15
Total	**100**
Malcolm Baldridge National Quality Award	
Leadership	120
Strategic planning	85
Customer focus	85
Measurement, analysis, and knowledge management	90
Workforce focus	85
Operations focus	85
Results	450
Total	**1000**
European Foundation for Quality Management	
Leadership	10%
People	10%
Strategy	10%
Partnerships and resources	10%
Processes, Products, and Services	10%
People results	10%
Customer results	15%
Society results	10%
Business results	15%
Total	**100%**

robust QM systems. In 2010, the programme's name was changed to the Baldrige Performance Excellence Program. The Baldrige Performance Excellence Program provides the Baldrige Excellence Framework, which contains the detailed criteria that describe a world-class total quality organisation. The programme and award are managed by the National Institute of Standards and Technology (NIST), an agency of the US Department of Commerce. The American Society for Quality (ASQ) administers the MBNQA under contract to NIST. It is claimed that each year up to 30,000 organisations request copies of the MBNQA criteria.

Up to three awards can be given each year across six categories: manufacturing, service company, small business, education, healthcare, and nonprofit. The applicants can be a whole firm or a legitimate business unit. The award is made by the President of the United States, with the recipients receiving a specially designed crystal trophy mounted with a gold-plated medallion. They may publicise and advertise their award, provided they agree to share information and best practices about their successful QM and improvement strategies with other American organisations.

Every Baldrige Award application is evaluated in seven major categories with a maximum total score of 1,000 (Baldrige Performance Excellence Program, 2024)). These are: leadership (115 points), strategy (90 points), customers (85 points), measurement, analysis, and knowledge management (90 points), workforce (85 points), operations (85 points), and results (450 points) see Table 11.2. The seven categories are subdivided into 17 items to address. It is claimed that 'In the most competitive business sectors, organisations with world-class business results are able to achieve a score above 700 on the 1,000-point Baldrige scale' (National Institute of Standards and Technology, 2025).

The criteria and processes are reviewed each year to ensure that they remain relevant and reflect current thinking, and, based on experience during the intervening period, their wording and relative scores are updated. It should be mentioned that, like the EFQM Model, the MBNQA criteria were developed originally by the business fraternity, management consultants, and academics (see www.nist.gov/baldridge/).

The EFQM Global Award

The EFQM Global Award is administered by the European Foundation for Quality Management (EFQM, see: efqm.org). Originally called the European Quality Award (EQA) this award was launched in October 1991 and first awarded in 1992. The award is assessed using the criteria of the EFQM excellence model. This was created by Europe's leading senior managers, academics, and consultants under the auspices of the EFQM and European Organisation for Quality (EOQ) and supported by the European Commission. According to the EFQM the EQA was intended to: 'focus attention on business excellence, provide a stimulus to companies and individuals to develop business improvement initiatives and demonstrate results achievable in all aspects of organisational activity'.

There are three forms of assessment that an organisation can use to evaluate its performance using the EFQM Model: EFQM Rapid Assessment, EFQM Advanced Assessment, and the EFQM Global Award. The EFQM Rapid Assessment is a streamlined, desk-based review intended to provide a high-level performance evaluation against a simplified version of the EFQM Model. The EFQM Advanced Assessment requires a detailed submission followed by a site visit conducted by a single assessor. The EFQM Global Award represents the highest level of recognition within the EFQM framework. It is awarded to organisations that exhibit outstanding performance and a sustained commitment to excellence as defined by the full EFQM Model. A central element of the award process is a site assessment conducted by a team of four to eight assessors over approximately one week. Recipients of the EFQM Global Award are granted a 'diamond' rating, signifying superior performance and positioning them as exemplars of excellence within their respective sectors.

The EFQM Model is assessed based on three interlinked guiding principles: Direction, Execution, and Results. Within these areas, there are seven criteria, which are connected in the widest sense, and also more explicitly, i.e.:

Criterion 1: Purpose, Vision, and Strategy
Criterion 2: Organisational Culture and Leadership

- Both linked to 'Direction'

Criterion 3: Engaging Stakeholders
Criterion 4: Creating Sustainable Value
Criterion 5: Driving Performance and Transformation

- Linked to 'Execution'

Criterion 6: Stakeholder Perceptions
Criterion 7: Strategic & Operational Performances

- Linked to 'Results'

The EFQM Model is based on what is termed the RADAR logic: results, approach, deployment, assessment, and review. The last four elements are used when assessing the enablers, and the result element is obviously used to assess results.

- Results cover what an organisation achieves and looks for: the existence of positive trends and sustained good performance, comparisons with previous, current, and future targets, comparison of results with competitors and best-in-class organisations, understanding the cause-and-effect relationships that prompt improvements, and ensuring that the scope of the results category covers all relevant areas.
- Approach covers what an organisation plans to do, along with its underlying reasons for this. It needs to be sound, systematic, appropriate, prevention-based, focused on relevant needs, and integrated with normal operations, and support organisational strategy.
- Deployment is the extent to which the approach has been systematically deployed and implemented throughout the organisation in all relevant areas.
- Assessment and Review relate to both approach and deployment. Progress will be subject to regular review cycles, analysis, and measurement, with appropriate learning and improvements planned, prioritised, and actioned.

The Shingo Prize

37 years after the creation of the Deming Prize in Japan, the Shingo Prize (shingo.org) was established in the USA, honouring the Japanese TQM industrialist Dr Shigeo Shingo. The Jon M. Huntsman School of Business at Utah State University established the award in 1988. Dr. Shingo is widely credited with developing TQM, just-in-time manufacturing, and making significant contributions to the Toyota Production System (TPS). The prize was created to promote awareness of lean manufacturing principles and to recognise organisations that successfully implement operational excellence. The award has evolved over time to recognise operational and organisational excellence across various sectors, including healthcare, government, education, and services.

The Shingo Model (Shingo Institute, 2025), which underpins the prize, is built upon a set of guiding principles organised into three dimensions: Cultural Enablers, Continuous Improvement, and Enterprise Alignment, and within these dimensions, there are ten guiding principles:

1 Respect Every Individual
2 Lead with Humility
3 Seek Perfection
4 Embrace Scientific Thinking
5 Focus on Process
6 Assure Quality at the Source
7 Improve Flow & Pull
8 Think Systemically
9 Create Constancy of Purpose
10 Create Value for the Customer

The Shingo Institute offers three levels of recognition: Bronze, Silver, and the Shingo Prize. Organisations typically progress through these levels as they mature in their operational excellence journey. The assessment process begins with an application based on their adherence to the Shingo Model principles and the results achieved. If the application is successful, a team of Shingo examiners conducts an on-site assessment. Unlike awards that often rely on a point system for specific criteria, the Shingo Prize employs a holistic assessment of both behaviour and results. The evaluation is primarily conducted against two key criteria: the Behaviour Assessment Scale and the Results Scale. The emphasis is on the interdependence of behaviours and results, rather than assigning a fixed percentage weighting between them. The award uses a 1,000-point system: the Shingo Prize is awarded for scores of 775 or higher, Silver Medallion for 675–774, and Bronze Medallion for 575–674.

Like the EFQM, the Shingo Prize has expanded internationally, with most Shingo Prize recipients now originating from outside North America. While the MBNQA, Deming, and EFQM have their creation around the fundamentals of QM, and evolved from there to excellence, the Shingo Prize was created around the principles derived from the TPS and lean thinking.

A key moment in the Shingo Prize and the use of self-assessments against awards is the bankruptcy of some recipient sites that had recently received the Shingo Prize, significantly affecting the credibility of operational excellence awards. This helped drive a significant change with the Shingo Prize and the EFQM to look at the future sustainability of organisations and a more holistic approach of QM and excellence culture.

The bankruptcy of some Shingo Prize recipients soon after receiving excellence awards raised concerns about the credibility of operational excellence awards. Since then, both the Shingo Prize and EFQM have been adapted to focus more on evaluating organisational sustainability and resilience.

The Self-Assessment Process

Self-assessment uses one of the models underpinning an award to pinpoint improvement opportunities and identify new ways in which to encourage organisational excellence. On the other hand, audits (often confused with self-assessment by less advanced organisations with respect to their development of QM) are carried out with respect to a quality system standard such as the ISO 9001 series and are, in the main, looking for non-compliance and assessing to see if the system and underlying procedures are being followed.

According to Dale et al. (2016) the five most important reasons for organisations starting self-assessment were to:

- Find opportunities for improvement
- Create a focus on a QM based on either the EFQM or MBNQA model criteria
- Direct the improvement process
- Provide new motivation for the improvement process
- Manage the business

This ranking provides a clear indication that internal issues are the most important motivation for organisations starting formal self-assessment. In some organisations a self-assessment process is introduced in response to changes in the operating environment, company direction, and competitors. The need for improvement is now recognised in most cases, regardless of how well an organisation may be doing.

After gaining the commitment of management to the self-assessment process and carrying out the necessary education and training, the following are the main steps that an organisation should follow in setting about self-assessment:

- Assess what the organisation has done well.
- Identify what aspects could be improved upon.
- Pinpoint gaps and what elements are missing.
- Develop an action plan to pick up the pace of the improvement process.

A key aspect of the process is taking an honest look at the organisation in order to identify its shortcomings; this process usually takes about three months.

There are several methods by which an organisation may undertake self-assessment, varying in complexity, rigour, resources, and the effort required. Each method has advantages and disadvantages, and an organisation must choose the one (s) most suited to its circumstances. Some organisations prefer to go for a full award simulation approach after using a matrix chart or questionnaire for educational purposes. Other organisations choose a more incremental approach. In general, organisations develop from using a simple approach to a more complex one, unless they have some external stimulus affecting the pace at which they address the process.

The choice of approach is dependent upon the level of QM maturity. Organisations with less experience should use the simpler methods, while those that are more advanced should adopt the more searching and rigorous methods.

Assessment against a model, whether by internal or external assessors, has three discrete phases:

1 Gathering data for each criterion.
2 Assessing the data gathered.
3 Developing plans and actions arising from the assessment and monitoring the progress and effectiveness of the plan of action.

The key issues that need to be considered by those organisations undertaking self-assessment for the first time include:

- Reach an agreement and be clear on the motivation for undertaking self-assessment and articulate the long-term expectations of the process.
- Senior management must be committed to the self-assessment process and be prepared to use the results to develop improvement plans.
- The people involved in the process need to be trained (i.e. training of assessors, data collectors, etc).
- Communicate within the business the reasons for and what is involved in self-assessment.
- Decide the self-assessment method (s) to be used
- Plan the means of collecting the data.

  – Decide the team and allocate roles and responsibilities for each criterion of the model
  – Develop a data-collection methodology and identify data sources
  – Agree on an activity schedule and manage it as a project.

- Decide the best way of organising the data that has been collected.
- Agree on the means of coordinating the process.
- Present the data, reach an agreement on strengths and areas for improvement, and agree on the scores for the criteria.
- Feedback to facilitate organisational learning.
- Prioritise the improvements, develop an action plan, and ensure that the ownership is with the appropriate people.
- Regularly review progress against the plan.
- Ensure that self-assessment is linked with the business planning process. This is important to ensure that the areas for improvement are turned into actions that are implemented.
- Repeat the self-assessment.

Success Factors for Self-Assessment

A key to the success of self-assessments is that they must be written down by the assessors. If they are purely verbal, they will have far less power. Something written down has a life of its own and can be referred to repeatedly. It can also be related to a business plan and a subsequent improvement plan, and can be part of a PDCA cycle. Written assessments also mean they will be taken much more seriously by both assessors and the assessed. Dale et al. (2016) have identified the following criteria, which are necessary for a successful self-assessment process:

- Gaining commitment and support from all levels of staff.
- Awareness of the use of the model as a measurement tool.
- Incorporation of self-assessment into the business planning process.
- Not allowing the process to be ‘added on’ to employees’ existing workload.
- Developing a framework for performance monitoring.

It is always difficult to clarify the impact of self-assessment on an organisation’s business results. However, those organisations with considerable experience in self-assessment do have a very positive perception of the impact of the techniques on their business performance.

Difficulties with Self-Assessment

There are a number of problems that hinder the self-assessment process: these are summarised below. These types of problems are created by a lack of senior management support, an inability to plan self-assessment effectively, and a low level of QM maturity:

1. Upfront costs of training – Many organisations underestimate the cost and time required to sufficiently train and develop their people in the Model and its assessment
2. Subjectivity in self-assessing – Teams may overrate themselves due to internal bias, overestimation, or lack of objectivity
3. Lack of External Validation – Without third-party audits or benchmarking, it’s hard to confirm the accuracy of the self-assessment
4. Misunderstanding of the Model – Teams may misinterpret the criteria or intent of external award frameworks and their models

5 Return on investment: Regular assessments can be time-consuming and costly if their insights aren't integrated into better decision making for management
6 Maintaining Self-Assessment: Teams fail to take ownership of regular assessments and see them as a 'tick box' activity rather than a valuable tool to help them improve

Summary

The self-assessment award frameworks described in this chapter can significantly enhance an organisation's pursuit of QM and improve organisational results. However, a crucial first step for any organisation is to honestly assess its current maturity in QM. If an organisation is primarily reactive with unstable and immature QM processes, top management should prioritise addressing core QM issues before pursuing an award model. That said, some frameworks are effectively promoted as roadmaps for start-up organisations, offering guidance on establishing a set of strong QM systems.

The fundamental purpose of self-assessment should always be clear to senior management: why are we doing this? The goal is to identify and close performance gaps, thereby delivering sustained organisational results. This is achieved through understanding the organisation's current QM status, defining the desired future state, and identifying the necessary actions and activities to bridge the gap between them. The process of training staff to assess the model, assembling data, preparing in-depth reports, and annual point scoring is an essential component of self-assessment. The trap organisations can fall into is to become focused on meeting the points criteria of an award. It's the execution of the action plans and solving the day-to-day quality problems and priorities that require senior management's focus in self-assessment. The benefit of using self-assessment against one of the recognised models is not the winning of the award but its adoption as a methodology to assess progress, using appropriate diagnostics, and identifying opportunities for improvement, not forgetting the need to satisfy and delight customers.

There is a substantial industry built around these awards and models, providing education and training, events, conferences, research, and consultancy services. While these can help organisations understand criteria, there's a risk. Often, internal 'champions' of award models emerge whose sole focus becomes securing the award, potentially losing sight of the core objective: improving organisational performance and results. When an award becomes seen as a cure-all rather than a means to an end, or when the focus shifts solely to scoring points against criteria, senior management can become obsessed with recognition instead of the crucial action plans identified by the self-assessment. A common pitfall for senior management is neglecting the execution of these action plans, prioritising the award itself over the day-to-day operational and QM activities that truly add value to customers and stakeholders.

Despite these potential pitfalls, a significant positive outcome of achieving a recognised award is the external validation and recognition of performance. This should not be underestimated, as it contributes to increasing employee engagement and a culture of continuous improvement.

Ultimately, while award frameworks can offer valuable guidance and recognition, an organisation's priority must be to ensure stable, existing systems and processes are in place. Engaging in an award process is often incredibly time-consuming and resource-intensive, and it should only be undertaken when the organisation is truly ready to leverage it for genuine performance improvement.

References

Baldrige Performance Excellence Programme (2024). *Baldrige: Performance Excellence Delivers World-Class Results*. National Institute of Standards and Technology (NIST). Available at: https://www.nist.gov/baldrige

BS EN ISO 9001 (2015). *Quality management systems: Fundamentals and vocabulary*. London: British Standards Institution.

College of Policing 2025. https://www.college.police.uk/support-forces/getting-organisational-support-your-peers

Dale, B., Bamford, D., & van der Wiele, T. (2016). *Managing Quality: an essential guide and resource gateway*, 6th edition (p. 330). UK: John Wiley & Sons.

Deming Prize Committee (2000). *The deming prize guide for overseas companies*. Tokyo: Union of Japanese Scientists and Engineers.

EFQM (2025). European Foundation for Quality Management. https://www.efqm.org

JUSE (2025). Union of Japanese Scientists and Engineers. https://www.juse.or.jp

National Institute of Standards and Technology (2025).

Shingo Institute. (2025). https://shingo.org/shingo-model/

12 Managing Quality

Challenges and Opportunities

Introduction

The discipline of quality management and organisational excellence is entering a period of transformation. This evolution is being driven by rapid technological progress, changing customer expectations, and increasingly complex global dynamics. For organisations committed to achieving excellence in quality, traditional models and frameworks are proving insufficient. This chapter examines the key themes and emerging challenges that quality management and organisational excellence initiatives must address to remain competitive, resilient, and relevant in the years ahead.

Future Challenges in Managing Quality and Organisational Excellence

Technological Advancements

Technological innovation will continue to exert a significant influence on quality management and organisational excellence over the coming decade. The core principles of the fourth industrial revolution—particularly digitalisation—have already disrupted

DOI: 10.4324/9781003594369-16

conventional organisational structures and operational models (Cranmer et al., 2022). The widespread adoption of advanced data analytics, the Internet of Things, artificial intelligence, cloud computing, and automation technologies is accelerating transformation across many sectors (tom Dieck et al., 2023). These developments are reshaping how organisations create value and manage operations, presenting both opportunities and challenges for the advancement of quality management. Organisations must now adapt their quality frameworks to accommodate these technologies, ensuring that systems are robust, data is reliable, and personnel are equipped to manage and improve processes in increasingly digital environments.

The integration of artificial intelligence and machine learning across organisational functions presents both significant opportunities and complex challenges for quality management. A central concern is the assurance of quality, reliability, transparency, and ethical behaviour within artificial intelligence systems themselves, alongside the effective use of these technologies to enhance traditional quality processes.

In operations, artificial intelligence has the potential to revolutionise quality control through predictive maintenance, automated defect detection, and optimised process methodologies. By analysing large volumes of sensor data from operational environments, artificial intelligence models can anticipate failures or quality deviations before they occur, enabling proactive interventions and reducing mistakes. This represents a shift from reactive to predictive quality control. However, poor implementation of artificial intelligence can result in biased outcomes, operational failures, and a loss of customer trust (Cranmer et al., 2022). Organisations must therefore develop new frameworks for artificial intelligence quality assurance and governance. Key to making this work is the establishment of structured, reliable data collection systems that ensure accuracy, completeness, and consistency.

Advanced Data Analytics and Predictive Quality

The increasing volume of data generated by operations necessitates more sophisticated analytical capabilities. The challenge lies in progressing beyond descriptive analytics—understanding what has happened—to predictive analytics, which anticipates future outcomes, and prescriptive analytics, which guides decision-making (Dehe et al., 2025). This evolution requires robust data infrastructure, skilled personnel with expertise in data science, and the ability to derive actionable insights from diverse data sources. Predictive quality management enables organisations to prevent defects, optimise processes, reduce waste, and enhance customer satisfaction. It shifts the emphasis from inspection to prevention, resulting in cost efficiencies and improved reputational standing (Bamford et al., 2023).

In the aerospace sector, for example, companies are employing advanced analytics to monitor jet engine performance—tracking variables such as temperature, pressure, and vibration—to predict component failures and schedule maintenance proactively (GE Aerospace, 2025). This approach enhances safety and operational reliability, which are critical dimensions of quality. In service delivery organisations, predictive analytics is being used to anticipate customer needs, identify potential service disruptions, and implement timely interventions, thereby improving the overall quality of the customer experience.

Measurement of performance and progress in meaningful terms is a difficult subject; however, it is vital to review the performance trend of the key measurables to ensure that the improvement initiatives and projects are having the desired impact (Bambrick et al., 2025). While in-house performance indicators are important, it is the customer assessment of the product and/or service which really counts, and it is important to have a set of measures which reflect the customer's viewpoint. There are a variety of means used to assess customer perceptions, obtain customer feedback, and understand the marketplace, including surveys, interviews, customer focus meetings, shows, launches, and even mystery shoppers. It is helpful to combine the data collected in this way into a customer satisfaction index.

The Human Dimension of Quality Management and Organisational Excellence

Despite our collective increasing reliance on technology, the human element remains central to the success of quality management and the cultivation of a culture of operational excellence. One of the most pressing challenges is the need to adapt the workforce to new technologies and evolving workplace dynamics. Digitalisation, automation, and artificial intelligence demand new skill sets, including digital literacy, analytical capability, and interpersonal competencies such as adaptability and effective communication (Mitchell et al., 2025). Organisations such as Bosch have recognised this imperative and invested significantly in workforce development, offering comprehensive training programmes in data analytics, the Internet of Things, and artificial intelligence. These initiatives enable employees not only to operate within highly automated environments but also to manage and continuously improve quality processes (Bosch, 2025).

Furthermore, increased workforce mobility introduces additional complexity. Ensuring data security across multiple locations, maintaining consistent adherence to processes without direct supervision, and developing a shared commitment to quality are all significant challenges. As teams become more geographically dispersed and face-to-face interaction reduces, organisations must adopt new approaches to training, collaboration, and oversight. These measures are essential to prevent quality deviations and to sustain a culture of excellence across varied work environments.

Lean and Continuous Improvement Throughout the Value Chain

Over the past three decades, Lean management and Continuous Improvement have become well-established practices within manufacturing and operational environments (Bamford et al., 2015). Rooted in principles of waste reduction and efficiency, Lean remains a cornerstone of quality management. However, a challenge lies in extending these methodologies beyond the confines of manufacturing, applying them effectively across entire organisations—including service sectors and public institutions.

Increasingly, Lean and Continuous Improvement are being integrated with digital transformation initiatives and advanced technologies. The principal challenge is to adapt Lean methodologies to fast-paced, iterative environments, leveraging data analytics and artificial intelligence to generate real-time insights and embed quality within continuous delivery cycles. This evolution marks a shift from reactive problem-solving to predictive quality management, where data-driven insights are used to anticipate and prevent issues before they arise. To support this transition, the development of workforce capabilities in digital literacy, analytical thinking, and problem-solving remains essential.

Cybersecurity and Data Privacy

As quality management systems become more digitised and interconnected—through technologies such as Internet of Things sensors, cloud-based platforms, and integrated supply chains—they are increasingly exposed to cybersecurity threats. Safeguarding the integrity, confidentiality, and availability of quality-related data is now a critical concern (Crossette-Thambiah et al., 2024). Moreover, evolving data privacy regulations demand careful handling of personal data within quality processes, particularly in areas such as customer feedback and product usage analytics. Quality professionals must therefore expand their remit to include advocacy for cybersecurity and data protection, ensuring that quality systems are not only effective but also secure and compliant.

Sustainability and Environmental Responsibility

Quality management is no longer solely concerned with product and service performance; it now encompasses environmental and social impact. Organisations are under growing pressure to design, produce, and manage offerings in ways that minimise environmental footprints and promote resource efficiency. This requires a lifecycle approach to quality—managing standards from raw material sourcing through to end-of-life disposal or recycling. Sustainability must become a central focus for quality professionals, with attention given to responsible sourcing, reduction of plastic waste, and the environmental impact of products and services. A notable example is Patagonia, the outdoor gear manufacturer, which has extended its quality management practices to include durability, repairability, and recyclability. By promoting a circular economy—encouraging customers to repair and reuse products and offering robust repair services—Patagonia ensures quality throughout an extended product lifecycle (Dezi et al., 2025).

Personalisation and Customer Experience

Customers increasingly expect personalised products and services, delivered through seamless and intuitive experiences across all stages of their interaction with an organisation (Oumaima & Lamari, 2024). Quality management must evolve beyond the traditional focus on product specifications to encompass the entire customer journey. This requires an authentic understanding of individual customer needs, preferences, and feedback—often in real time—and the ability to translate these insights into meaningful quality improvements. Operational excellence must be matched by a high-quality customer experience, with quality management systems designed to support and enhance every aspect of the customer's engagement with the organisation.

Supply Chain Resilience and Transparency

Quality management faces growing challenges in supplier development and oversight, particularly as supply chains become more vulnerable to disruption from environmental, geopolitical, and economic factors. Over the past thirty years, global supply chains have become increasingly complex and interdependent. However, the first two decades of the twenty-first century have seen significant disruptions, prompting organisations to reassess their supply chain strategies (Heinis et al., 2022). Quality management must now ensure consistent standards across intricate supplier networks—from raw material providers to final distribution channels. This demands unprecedented levels of transparency, traceability, and resilience. Managing supplier quality, mitigating risks, and responding swiftly to disruptions will remain persistent and critical challenges.

Regulatory Compliance and Global Standards

Organisations operating across multiple geographical territories must navigate an expanding array of regulatory requirements and quality standards, many of which are diverse and occasionally conflicting. Maintaining compliance, managing certifications, and adapting quality systems to meet these varied demands is a continuous and complex undertaking (Papalexi et al., 2021). This challenge is particularly acute in highly regulated sectors such as healthcare, automotive, and aerospace, where adherence to stringent standards is essential. Quality professionals must remain vigilant and well-informed, ensuring that systems are both compliant and adaptable to evolving regulatory landscapes.

Framework: Future Challenges in Quality & Organisational Excellence

This framework summarises the key future challenges in managing quality and organisational excellence, along with suggested actions to address each challenge. It is designed to support strategic planning, operational alignment, and workforce development across sectors (Table 12.1).

Opportunities in Managing Quality and Organisational Excellence

Strategic Leadership and Managing Quality?

The adoption of advanced technologies within organisations employing fewer than 500 people remains largely embryonic. These smaller enterprises, which constitute the majority of businesses globally, have yet to embrace digital transformation in any meaningful way. In contrast, larger corporations—typically early adopters—have begun to integrate these technologies more systematically (Cranmer et al., 2022).

Fundamentally, this is a matter of mindset, and it is one that must be driven from the top. Unfortunately, in many Western organisations, leadership tends to operate within a short-term investment framework. This stands in stark contrast to the long-term strategic thinking observed in countries such as Japan, China, and Germany. Historically, Japan offers a compelling case study. Following the Second World War, Japanese products were widely perceived as cheap and of poor quality. However, by the 1980s, Japan had become synonymous with high-quality manufacturing, particularly in electronics. This transformation was not accidental—it was the result of a national commitment to quality, inspired by the teachings of Western quality experts such as Deming and Juran, whose philosophies were embraced and embedded into Japanese corporate culture (Dale et al., 2016). China followed a similar trajectory. 40 years ago, products labelled "Made in China" or "Made in Taiwan" were often associated with low cost and low quality. Today, however, China has evolved into a global leader in manufacturing quality, largely due to its cultural predisposition to follow top-down directives and long-term planning.

Western companies, particularly those in the UK and the US, tend to prioritise short-term financial metrics such as EBITDA (earnings before interest, taxes, depreciation, and amortisation) according to Olbert (2025). This focus on rapid returns—typically within 12 to 24 months—limits their ability to invest in long-term quality improvement initiatives. German firms, especially the Mittelstand companies (small and medium-sized enterprises that are the backbone of the German economy; they are often family-owned, prioritise

Table 12.1 Summary of future challenges, key issues, and suggested action

Challenge Area	*Key Issues*	*Suggested Actions*
Technological Advancements	Disruption of traditional models via digitalisation, AI, IoT, and automation	– Update quality frameworks to integrate digital tools – Ensure data reliability and system robustness – Upskill workforce for digital environments
Advanced Data Analytics & Predictive Quality	Shift from descriptive to predictive/prescriptive analytics	– Invest in data infrastructure and analytics platforms – Develop data science capabilities – Use predictive insights to prevent defects and optimise processes
Human Dimension	Workforce adaptation to tech-driven environments	– Provide training in digital literacy, analytics, and communication – Foster a culture of adaptability and continuous learning – Empower employees to lead quality improvements
Lean & Continuous Improvement Across Value Chain	Limited application beyond manufacturing	– Extend Lean principles to services and public sectors – Promote cross-functional CI initiatives – Align Lean with digital transformation efforts
Cybersecurity & Data Privacy	Increased exposure to cyber threats and regulatory demands	– Embed cybersecurity into quality systems – Train quality professionals in data protection – Ensure compliance with privacy regulations
Sustainability & Environmental Responsibility	Quality now includes environmental and social impact	– Adopt lifecycle quality management – Promote circular economy practices – Integrate sustainability metrics into quality KPIs
Personalisation & Customer Experience	Rising expectations for tailored, seamless experiences	– Use real-time customer data for quality improvements – Design systems around customer journey touchpoints – Align operational excellence with customer experience excellence
Supply Chain Resilience & Transparency	Vulnerability to disruptions and complexity	– Enhance supplier quality management – Build traceability and transparency mechanisms – Develop agile response strategies for disruptions
Regulatory Compliance & Global Standards	Diverse and evolving global regulations	– Monitor and adapt to changing standards – Maintain flexible and scalable compliance systems – Ensure cross-border quality consistency

long-term success, and are known for their innovation and strong market position), offer a notable exception (Bebic et al., 2025). Their strategic horizon often spans 20 to 25 years, enabling them to embed quality deeply into their operations. In contrast, the median tenure of a strategic leader, such as a Chief Executive Officer (CEO) in Western firms, is approximately 4.8 years (Chen, 2023). This limited timeframe makes it exceedingly difficult to effect meaningful cultural or behavioural change. As a result, the potential for competitive advantage through quality management remains largely misunderstood and underutilised. It is often treated as a standalone function rather than a company-wide philosophy.

Additionally, many organisations have adopted the tools and techniques pioneered in Japan and relatively well known in the West—such as Lean, Kaizen, and Six Sigma—but they fail to adopt the underlying mindset. One which requires active strategic leadership, top-down direction, clear two-way communication, fulfilment of promises, et cetera. Consequently, quality professionals are frequently relegated to roles focused on compliance and assurance, rather than being empowered to drive proactive cultural transformation. Their true value lies in educating and developing others across the business, fostering a shared understanding of quality as a strategic imperative (Nadeem, 2023).

Compounding these challenges is the rapidly evolving geopolitical landscape. Many companies have recently withdrawn from the US market, while others are actively investing in it (Sternfels et al., 2025). The pace of change—who we do business with, how we operate, and what we sell—has accelerated dramatically. Quality management must adapt to this new reality. It can no longer be a reactive function; it must become a proactive enabler of strategic agility. In summary, the future of quality management depends not on tools or frameworks alone, but on a fundamental shift in mindset—one that must be led from the top and sustained over the long term. Only then can organisations truly realise the value of quality as a driver of competitive advantage. But how should organisations set themselves up for operational and quality excellence?

Figure 12.1 presents a conceptual representation of the need and opportunity to align managing quality and organisational excellence. From the extensive research and consultancy of the authors, we believe that approximately 5% of companies are currently able to adopt technology appropriately, proactively use quality management for competitive advantage and clarity of communication, and also consider the need to both motivate and inspire their people. This ambitious realisation of potential is possible but requires planning, commitment, skill, and motivation—initially from the top down within organisations, and then from the middle out and the bottom up.

Quality management is no longer what it once was. The landscape is shifting rapidly, driven by digitalisation, artificial intelligence, and rising customer expectations. The old frameworks, while still valuable, need to be rethought and reshaped to meet the demands of today's fast-moving, interconnected world. Lean and Continuous Improvement remain central, but their reach must now extend beyond operations. Service sectors, public

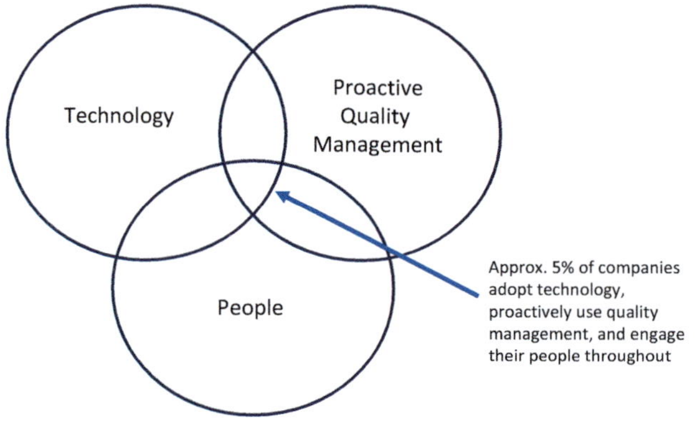

Figure 12.1 How to align managing quality and organisational excellence?

institutions, and digital environments all require agile, data-informed approaches to quality. The fusion of traditional methods with advanced analytics and real-time feedback is transforming how organisations operate, solve problems, and deliver value. At the same time, the human element cannot be overlooked. As systems become more automated and dispersed, the workforce must be equipped with new skills—digital literacy, analytical thinking, and the ability to collaborate across boundaries. Cybersecurity and data privacy are now part of the quality conversation, as interconnected systems bring new risks and responsibilities. Sustainability is also moving up the agenda. Quality is no longer just about performance—it is about environmental impact, responsible sourcing, and lifecycle thinking. Customers expect more than good products; they want ethical practices and seamless, personalised experiences. And with supply chains under pressure from global disruptions, resilience and transparency have become essential. Navigating this complexity requires not just compliance with global standards, but a proactive, forward-thinking approach to quality—one that sees change not as a threat, but as an opportunity.

We perceive four specific opportunities for business:

1 Digital Quality Transformation: Businesses can harness artificial intelligence and predictive analytics to move from reactive quality control to proactive, data-driven decision-making—reducing waste, improving reliability, and enhancing customer satisfaction.

Table 12.2 Summary of opportunity areas, key issues, and suggested action

Opportunity area	*Key issues*	*Suggested actions*
Digital Quality Transformation	– Limited adoption of advanced technologies in SMEs – Reactive rather than proactive quality management – Lack of data-driven decision-making	– Invest in AI and predictive analytics for quality control – Train leaders and staff in digital literacy – Develop real-time quality dashboards and KPIs
Service Sector Lean Integration	– Lean tools are often misapplied or misunderstood – Service sectors overlook waste and variation – Quality seen as compliance, not strategic	– Apply Lean principles to service and admin functions – Educate teams on Lean thinking and continuous improvement – Empower quality professionals to lead transformation
Sustainable Product Lifecycle Management	– Short-termism limits investment in sustainability – Lack of integration between quality and environmental goals – Consumer demand for sustainable products is rising	– Embed sustainability into quality systems – Focus on durability, repairability, and circular economy – Align product development with environmental, social, and governance (ESG) goals
Resilient and Transparent Supply Chains	– Geopolitical instability affects supply chain reliability – Poor supplier development and traceability – Reactive risk management practices	– Invest in supplier partnerships and traceability tech – Develop robust risk management frameworks – Use quality as a strategic lever for supply chain resilience

Table 12.3 Summary of crosscutting strategic leadership issues, challenges, implications, and strategic response

Challenge	*Implication*	*Strategic response*
Short-Term Leadership Tenure	Limits cultural and behavioural change	Extend strategic planning horizons; embed quality into long-term vision
Mindset Misalignment	Tools adopted without philosophy	Promote top-down commitment to quality culture; integrate quality into leadership development
Western vs. Eastern Strategic Thinking	Western firms focus on EBITDA; Eastern firms on long-term excellence	Benchmark against long-term models (e.g., Japan, Germany); shift focus from compliance to competitive advantage
Underutilisation of Quality Professionals	Quality is seen as assurance, not transformation	Reposition quality roles as educators and enablers; foster two-way communication and empowerment

2 Service Sector Lean Integration: Applying Lean principles to service delivery and administrative functions offers significant efficiency gains, particularly in healthcare, education, and public services, where waste and variation are often overlooked.
3 Sustainable Product Lifecycle Management: Organisations can differentiate themselves by embedding sustainability into their quality systems—focusing on durability, repairability, and circular economy models that appeal to environmentally conscious consumers.
4 Resilient and Transparent Supply Chains: Investing in supplier development, traceability technologies, and risk management frameworks enables businesses to maintain quality standards even in the face of geopolitical or environmental disruption.

Framework: Opportunities in Managing Quality and Organisational Excellence

Please see below a summary of what we have identified as crosscutting strategic leadership issues (Table 12.3).

Summary

Final reflections? In the pursuit of organisational excellence, it is often the operational fundamentals that are most neglected. Time and again, senior leaders—supported by internal teams and external consultants—overlook the core principles of managing quality: clarity of purpose, consistency in execution, and a deep respect for people, processes, and systems. Misjudgements around technology, communication, training, infrastructure, and measurement are not just common—they are costly. Too often, organisations are drawn to fashionable methodologies or popular frameworks, losing sight of what truly drives sustainable performance.

Excellence is not elusive. It begins with a commitment to mastering the basics and embedding quality into the fabric of everyday operations. With sound knowledge, strategic intent, and disciplined planning, organisations can avoid these pitfalls and build environments where quality is not an initiative, but a way of working. If this book has helped you take even one step towards that goal—towards a more thoughtful, effective, and resilient approach to managing quality—then it has done its job.

References

Bambrick, S., Vafadarnikjoo, A., & Reid, I., Bamford, D. (2025). Unpacking critical success factors to improve supply chain effectiveness, efficiency and performance: A 7Vs framework for consideration. *Production Planning & Control*, 36(4), 521–535.

Bamford, D., Forrester, P., Dehe, B., & Leese, R.G. (2015). Partial and iterative lean implementation: Two case studies. *International Journal of Operations & Production Management*, 35(5), 702–727.

Bamford, D., Forrester, P., & Reid, I. (2023). *Essential guide to operations management: Concepts & case notes*, 2nd edition. UK: Routledge.

Bebic, M., Badie, N.B., Tyll, L., & Srivastava, M. (2025). Exploring the barriers and drivers of ESG in the German Mittelstand: A qualitative analysis of mechanical and plant engineering companies. *Corporate Social Responsibility and Environmental Management*, 32(2), 2147–2170.

Bosch (2025) *Bosch uses software and AI to make its product smarter and make people live safer*, https://us.bosch-press.com/pressportal/us/en/press-release-26240.html, accessed August 2025.

Chen, J. (2023). *CEO tenure rates*. Harvard Law school forum on corporate governance, https://corpgov.law.harvard.edu/2023/08/04/ceo-tenure-rates-2/, accessed August 2025.

Cranmer, E., tom Dieck, M., Papalexi, M., & Bamford, D. (2022). Internet of things: Aspiration, implementation and contribution. *Journal of Business Research*, 139, 69–80.

Crossette-Thambiah, G., Berleant, D., & AbuHalimeh, A. (2024). An information quality framework for managed health care. *Journal of Healthcare Leadership*, 16, 343–364.

Dale, B., Bamford, D., & van der Wiele, T. (2016). *Managing Quality: An essential guide and resource gateway*, 6th edition. UK: John Wiley & Sons.

Dehe, B., Subasinghage, M., & Mirzaei, M. (2025). Orchestrating resources for big data analytics implementation in manufacturing SMEs: Insights into managerial role and engagement. *Production Planning & Control*, 1–33.

Dezi, L., Hysa, X., Calabrese, M., & Mercuri, F. (2025). Open total quality management in the circular economy age: A social enterprise perspective through the case of Patagonia. *Total Quality Management & Business Excellence*, 36(3–4), 309–323.

GE Aerospace (2025) *Every flight counts – using data to improve engine performance*, https://www.geaerospace.com/news/articles/people-product-technology/every-flight-counts-using-data-improve-aircraft-engine, accessed August2025.

Heinis, S., Bamford, D., Papalexi, M., & Vafadar, A. (2022). Services procurement: A systematic literature review of practices and challenges. *International Journal of Management Reviews*, 24(3), 352–372.

Mitchell, T., Bamford, D., Papalexi, M., Reid, I., & Webster, R. (2025). Impact pathways: High-performance sports operationalisation through Industry 4.0: An actionable blueprint. *International Journal of Operations and Production Management*, 45(13), 253–265.

Nadeem, M. (2023). Leadership's role in effective quality assurance implementation in higher education institutions. *Education Quarterly Reviews*, 6(3), 237.

Olbert, L. (2025). Financial analysts' use of industry specific stock valuation models. *Journal of Applied Accounting Research*, 26(6), 108–138.

Oumaima, J., & Lamari, S. (2024). Customer experience in the digital transformation era: Insights on personalization, digital marketing, and customer relationship management. *International Journal of Economics, Management and Finance (IJEMF)*, 3(2), 52–69.

Papalexi, M., Bamford, D., Nikitas, A., Breen, L., & Tipi, N. (2021). Pharmaceutical supply chains and management innovation. *Supply Chain Management: An International Journal*, 27(4), 485–508.

Sternfels, B., Singhal, S., Levy, C., Weddle, B., Fox, Z., & Watters, M. (2025). Multinationals at a crossroads: Adapting to a new geopolitical era. *The McKinsey Quarterly*. http://www.mckinsey.com.https://www.mckinsey.com/capabilities/geopolitics/our-insights/multinationals-at-a-crossroads-adapting-to-a-new-geopolitical-era

tom Dieck, M., Cranmer, E., Prim, A., & Bamford, D. (2023). Can Augmented Reality (AR) applications enhance students' experiences? Gratifications, engagement and learning styles. *Information Technology & People*, 37(3), 1251–1278.

Index

Please note that page numbers in **bold** denote tables, page numbers in ***bold/italics*** denotes boxes, while those in *italics* refer to a figure.